Copyright© 2020 art of motion training in movement gmbh.

All rights reserved. No part of this publication may be reproduced, distributed, or transmitted in any form or by any means, including photocopying, recording, or other electronic or mechanical methods, without the prior written permission from art of motion training in movement® gmbh, except in the case of brief quotations embodied in critical reviews.

For permission requests, write to art of motion training in movement® gmbh, at welcome@art-of-motion.com.
Thank you for your respect!

1. Edition: August 2008
10. Edition: February 2020
1. Publication: June 2020
ISBN: 978-1708339302
Self-publishing: Amazon Media EU S.à r.l., 5 Rue Plaetis, L-2338, Luxembourg

Author: Karin Gurtner
Editors: Kiki Vance and Heidi Savage

Graphic design: Babuche Gruber – TanzArt, www.tanzart-bern.ch
Photography: Felix Peter – Fotogigant, www.fotogigant.ch
Photo contribution: Anita Preece Kopp and Richard Preece – origin8, www.origin8.ch

Personal Responsibility
It is always the responsibility of the individual to assess his or her own physical constitution and wellbeing before participating in any training activity. Whilst every effort has been made to ensure the content of this book is as technically accurate as possible, neither the author nor the editors or contributors assume any liability for any injury and/or damage to persons.

SLINGS
A DELIBERATE, FASCIA-FOCUSSED MOVEMENT PRACTICE

What moves you? Muscles, ambition, a desire for change or consistency, the need for safety or adventure, the wish to strengthen your health or integrity? All of it is influenced and empowered by fascia, the adaptable, collagenous connective tissue web inside your body.

What you recognise in the mirror as 'you' is shaped by the fascial system. Its architecture is a reflection of the way you stand, move, think, and feel. Because fascia and movement are intrinsically linked, a deliberate, fascia-focussed movement practice is a powerful medium to become and empower the best version of yourself. The Slings concept provides you with experience-based, science-informed tools to refine your postural balance, enhance your movement ease, strengthen trust in your body's resourcefulness, and invigorate your vitality.

This is going to be an amazing journey of 'wonderstanding' fascia in motion. This will also be a lifelong journey. Therefore, we had better get started!
Warmly,

CONTENT

Foreword By Thomas W. Myers	6
Introduction	**16**
Slings Practice	19
Fascial System in a Nutshell	22
Fascial Architecture	26
Somatic Fascia Differentiated	40
Skin to Bone Layering Model	44
Part 1: Slings Myofascial Training Concept	**50**
Resource-Oriented Integrative Movement	54
6 Guiding Principles	58
Part 2: Slings Myofascial Training Trinity	**62**
12 Fascial Movement Qualities	66
1. Tensile Strength	73
2. Muscle Collaboration	93
3. Force Transmission	105
4. Adaptability	119
5. Multidimensionality	139
6. Fluidity	153
7. Glide	167
8. Elasticity	183
9. Plasticity	207
10. Tone Regulation	221
11. Kinaesthesia	251
12. Imponderability	287
12 Slings Myofascial Training Techniques	288
12 Slings Training Aims	312
In Sum Total: 6 Good Reasons For Slings	319

Part 3: Slings Myofascial Training Applications   **322**

8 Slings in Motion Teaching Principles   325

    1. Fascial Movement Qualities   326

    2. Myofascial Meridians   326

    3. Differentiated Integration Training   328

    4. Functional Choreography   330

    5. Flow   337

    6. (Re)Balancing on the Go   340

    7. Resource-Oriented, Fascia-Focussed Communication   343

    8. Authenticity   347

Slings Lesson Planning Guide   350

Part 4: Parts of the Whole and the Whole of the Parts   **354**

4 Fascial Types   357

    Superficial Fascia   357

    Loose Fascia   360

    Deep Fascia   363

    Muscle Fascia   366

Movement as a Synergy   372

The Problem with Holism   376

Movement is Greater Than the Sum of Its Parts   380

Afterword   382

References   387

Tensegral Collaboration   391

More Good Things to Feed Your Brain and Nourish Your Body   392

FOREWORD

Relax. This is more than a beautiful book; it is an important piece of work about our inner architecture – both its science and its perception. Take a breath before you peek inside. Prepare yourself for its scope, far more applicable than just another yogilates sequence. Now open the book and open yourself to the patient workings of Karin Gurtner's "bodymind".

This presentation is – as Karin insists movement is – greater than the sum of its parts. I suggest dropping in at random at first to get a sense of Karin's voice. Take in the stunning photographs, the careful explanatory detail, and the trans-migrational knowledge emanating from any section you happen to pick. Then begin at the beginning, settling into this steel-hand-in-a-velvet-glove of a teacher who both lives her work and lives to communicate it to others.

When Karin Gurtner first showed up as a student in my advanced manual therapy training, I had no idea she had never done hands-on work of this type before. Seemingly shy, she submitted humbly to our guidance, progressed rapidly, clearly had no trouble shifting concepts into practice – but my focus was on the rest of the students in that class. A facilitator's attention is so often drawn by those with more difficulty.

Only later did I learn what a powerhouse had been in my training. Karin already had a well-developed school of Contemporary Pilates in Switzerland. In retrospect, I commend her willingness to enter this new arena with innocence. But, as it turned out, she had a well-disguised hidden agenda. Over the ensuing years, Karin wove the Anatomy Trains map and fascial properties exposed by research into a tapestry of the contemporary movement practice she calls Slings.

Now that Slings has blossomed, we have taught together, and she has issued me a challenge. In that original training, I conveyed to her my version of Anatomy Trains Structural Integration, the protocol of manual fascial release I developed from what Ida Rolf taught me. Now Karin avers that she can get comparable results – integration of the structure – from this program of movement.

In class, I have experienced the power and precision of her approach and felt its benefits. If you follow the step-by-step program outlined in the movement sequencing, you will arrive at a happy and adaptable intersection among the best of contemporary yoga, Pilates, and personal training practices. Deceptively simple, the

Slings program offers strengthening, elasticising, and coordinating in a way no manual therapy regime – not even mine – could ever provide. Slings also shares elements otherwise unique to Structural Integration – glide within and among layers, and the development of even tone for resilient, lengthened balance in 'acture' – posture in motion.

This book is the latest – and most daunting, from my ego's point-of-view – expression of this productive 'argument' in which Karin and I engage. Perhaps someday we can set up an experiment that would serve to settle the question, but in truth, both of us have been more helped by the mutual questioning than arriving at any answer. It helps winnow the chaff of outmoded ideas from the wheat of what truly works for whom and when.

As we have engaged in this give-and-take, I have gotten to know Karin better, so let me embarrass her a little as I close. You will see immediately that you can rest confidently in the precision in her exercises, sequences, and rationales because that is her Swiss discipline: whatever comes out for public consumption must be tested and exact.

But do not miss the deeply feeling, sensitive, passionate, sharing nature of the person within. This is not work born of the balletic abstraction of a studio and a Reformer; this is work born of troubles shouldered, impediments mastered, and long-held questions lived through to their answers. In short, for sure, she is a model for movement – I mean, look at those photos – but she is also a dedicated mensch as a teacher, a hand-me-the-shovel worker, and a friend.

This program is not only designed to help you look and feel good; it is designed to build the inner strength to see you through both your personal challenges and our 'interesting times'.

Tom Myers
Clarks Cove, Maine USA
April 2020

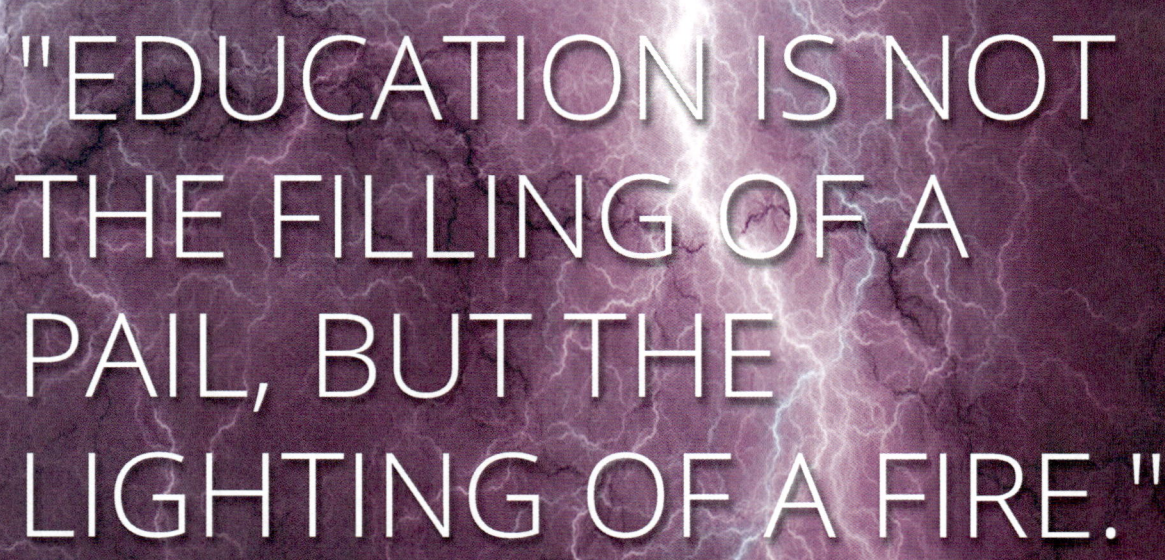

> "EDUCATION IS NOT THE FILLING OF A PAIL, BUT THE LIGHTING OF A FIRE."
>
> PLUTARCH

INSPIRED LEARNING

Clear Intention and Quality of Movement

For a number of years, I have continuously merged and translated experience and scientific information into practical application. Through this ongoing learning process, I have come to understand the value of twelve fascial qualities relevant to movement. From there, I developed a set of training techniques to achieve specific short-term and long-term training aims such as postural balance, movement ease, vitality, and a sense of being at home in your body. Because the concept works so well in real life, I wrote this book for curious movers and open-minded movement professionals who are interested in deliberately using fascia as a medium for positive change.

It is my aspiration to create an engaging learning experience in which you will:

- Learn new things about your body and what moves you
- Deepen and refine existing knowledge
- View things that you previously learned in a different way
- Understand what you feel from a different perspective
- Develop new strategies to strengthen your wellbeing holistically
- Enhance your movement vocabulary

A Unique Composition

As much as I would love to say that everything I am presenting to you is entirely original, the truth is, it is not. However, here is what I can offer you that is original:

- An intelligent fascia-focussed movement concept that is unique in its composition
- An experience-proven bridge from current scientific information to practical application
- Practical applications that are coherent with the theory presented
- A versatile spectrum of new and known exercises with a new fascial dimension
- A holistic concept that is inclusive and applicable to a wide variety of movement methods
- A complex, fascia-focussed movement puzzle, presented in a tangible and digestible manner

Appreciating Your Knowledge and Experiences

In whatever is outlined in this book, there is always room for your own thoughts, interpretations, and contributions.

WITH GRATITUDE AND HUMBLENESS

Despite dedicating a good number of years to the creation of a Contemporary Pilates curriculum, building a school, and travelling the world as an educator, developing the Slings Myofascial Training® concept and writing this book have been among the most defining, challenging, satisfying, and humbling experiences of my life thus far. Defining, because it helped me cognitively understand somatic experiences to create a relatable concept. Challenging, because of what I had to leave out and the insufficiency of my words to describe the richness of the 'real thing'. Deeply satisfying, because of the benefits it brings to others – and myself. Humbling, because I am confronted with my mistakes regularly, which keeps me grounded and advancing.

To You, Dear Reader

Thank you, the reader, for your embrace of the best story I am able to tell you at this moment in time. I do hope that you get as much out of this book as I learned from writing it.

To Those Referenced

Much gratitude goes to the scientists and researchers who inspired this work and provided the theoretical foundation for the practical applications. If you are among the brilliant minds listed at the end of this book, thank you for your contribution to the Slings concept, and thank you for your tolerance of my interpretation of your insights.

To Mone, Marion, Martina, Muriel, Gemma, Kurt, Kiki, Heidi, Babuche, and Felix

Abundant gratitude goes to my twin sister Mone for making everything I do possible, including this book. To my mother, Marion, and my close friends Martina, Muriel, and Gemma for seeing the potential of this work before I did. To my father, Kurt, for trusting me. To my English editors Kiki and Heidi, for their input and unwavering encouragement to get the word out. To graphic designer extraordinaire Babuche, who pays attention to every detail, again, and again. To Felix, fellow world-traveller and world-class photographer, who never shies away from a photo adventure in grand or odd locations. Without all of you, I wouldn't have had the willpower to keep (re)writing or the courage to finally publish this book.

To the Special People of the art of motion Team

A heartfelt thank you to everyone at art of motion for your incredible commitment, your adaptability, deeply embodied belief in the value of this work, and the skill and creativity with which you share it with others.

To Thomas W. Myers

Thank you, Tom, for inspiring my work and this book with the Anatomy Trains concept, your approach to structural integration, our collaboration, insightful conversations on Hawai'ian islands, and cobblestone streets in Spain, and of course, your being.

Karin feeling grateful at the Avenue of the Stars in Hong Kong in China

HOW TO USE THIS BOOK FOR INTELLECTUAL UNDERSTANDING AND EMBODIMENT

The content of this book contains food for thought and nourishment for the senses in the form of text and artistic photographs.
Text, because it is an effective way to convey ideas and knowledge.
Photographs, because an image can say more than a thousand words, create a lasting impression, and amplify the meaning of text.

Match Your Reading Style and Intention

When you pick up a non-fiction book, you can read it simply to grasp the main point of the topic or to fully understand the details of a subject you are exploring. There is no right or wrong way of doing it. However, it is useful to align your reading style with your intention.

To Readers Without a Specific Agenda
If you are reading this book simply because the topic is of interest to you, I hope you enjoy the content. I have written it as much for you as for the reader on an educational mission.

To Readers on a Mission
If you are reading this book with the intent to deeply understand the movement-related implications of the science-informed details, you are definitely on an educational mission! Regardless if you aspire to gain personal growth or professional development, to get the most out of this book I recommend a three-step process:

Step 1: Read the whole book from front to back without an agenda.

Step 2: Read it a second time with your educational mission in mind. Read it chapter by chapter. After each section, pause to reflect. Ask yourself if you have truly understood what you have just read.

Step 3: From memory, take time to summarise the most important things from the chapter you have just read. Compare notes and add forgotten valuable pieces to your summary.
After finishing, re-read your summaries and feel a sense of achievement for the knowledge you have attained.

EMBODY YOUR UNDERSTANDING WITH VIDEO PRACTICES

Included in this book are references to complimentary Slings exercise sequences. Additionally, there are links to an educational online video library that may be purchased. The video library contains twelve practices, one for each of the twelve Fascial Movement Qualities. Non-anatomical short stories accompany the practices. All of the material is designed to assist your embodiment and 'embrainment' of Slings Myofascial Training in an enjoyable, and I believe effective manner.

This movie-roll icon depicted to the right, when displayed in the book, indicates that there are relevant stories and practices available in the Slings Essentials video library or on the art of motion Academy YouTube channel.

Knowledge is power
Embodiment is power
Embodied cognition is a superpower

Slings Essentials Video Library on Vimeo on Demand

If you would like to treat yourself to the Slings Essentials Embodiment practice library, we will treat you to a 25% discount.

The video library is available on Vimeo on Demand:
vimeo.com/ondemand/slingsessentials

Discount code for a 25% price reduction:
Slings-Essentials-VIP

Complimentary Slings Movement Sequences on YouTube

References for the free exercises are found at the end of the Slings Practice chapter on page 21.

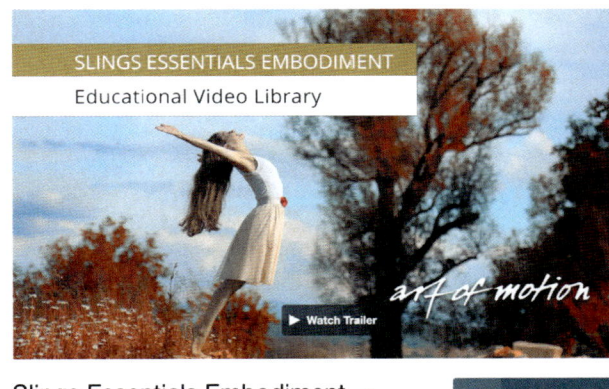
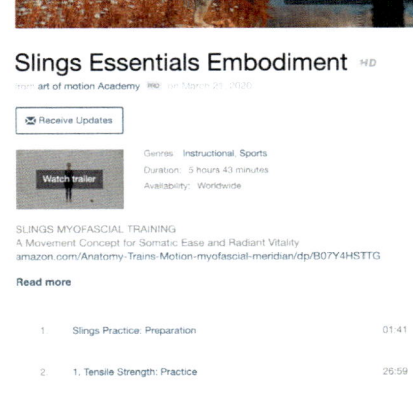

13

Concept and Book with Tensegral Qualities

"Easy learning does not build strong skills."   Brené Brown

The Slings concept and this book have a tensegral structure. Let me explain what I mean.

Non-linear learning requires persistence and trust

In a training method, an education concept, or a book that has a linear structure, you can begin with topic A, master it or understand it, then move on to topic B, and so forth. This isn't possible with a non-linear theory. Although there is structure, its organisation may not be initially recognisable. As a reader, mover, and learner taking in information in this non-linear way requires a healthy degree of persistence and trust in the process. It may also be necessary to put questions and uncertainties to the side while progressing. Practically speaking, this means that at the beginning of this book, you might read something that isn't entirely clear because it is explained in the context of one of the twelve Fascial Movement Qualities later on. The same goes for the practical examples. It is my intent to guide you through the material in a structured manner that allows the individual elements to gradually assemble, forming a coherent, yet tensegral whole.

Navigating This Book

Delta, the fourth letter in the Greek alphabet, looks like a triangle It means directed-trajectory. The connotation of 'direct' includes 'to guide an action' or 'point to a specific aspect'. 'Traject' stands for 'conveying from one person or place to another'. In that sense, a directed-trajectory is guided conveyance, which makes the triangle a matching symbol for the four main parts of this book.

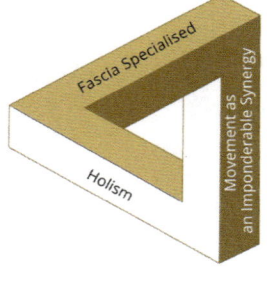

Introduction
- Slings Practice
- Fascial System in a Nutshell

Part 1: Slings Myofascial Training Concept
- Resource-Oriented Integrative Movement
- Deliberate and Fascia-Focussed Training
- 6 Guiding Principles

Part 2: Slings Myofascial Training Trinity
- 12 Fascial Movement Qualities
- 12 Myofascial Training Techniques
- 12 Slings Training Aims

Part 3: Slings Myofascial Training Applications
- 8 Teaching Principles
- Lesson Planning Guide
- Training Guidelines

Part 4: Parts of the Whole and the Whole of the Parts
- Fascia Specialised
- Movement as a Synergy
- Holism

The content of the book is interrelated; therefore, learning is non-linear. Feel free to flow between parts and chapters in whatever way is most useful to you. Let the fun begin!

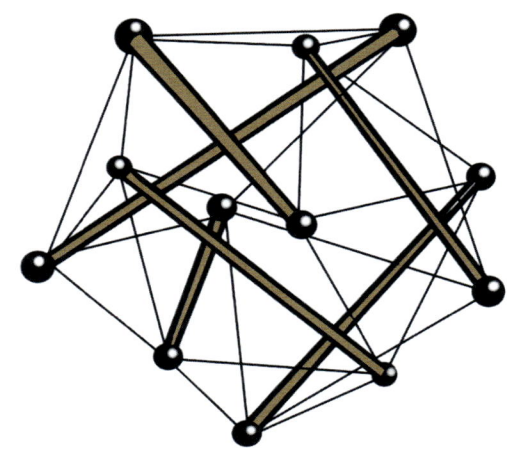

INTRODUCTION

SLINGS PRACTICE
FASCIAL SYSTEM IN A NUTSHELL

PRACTICE

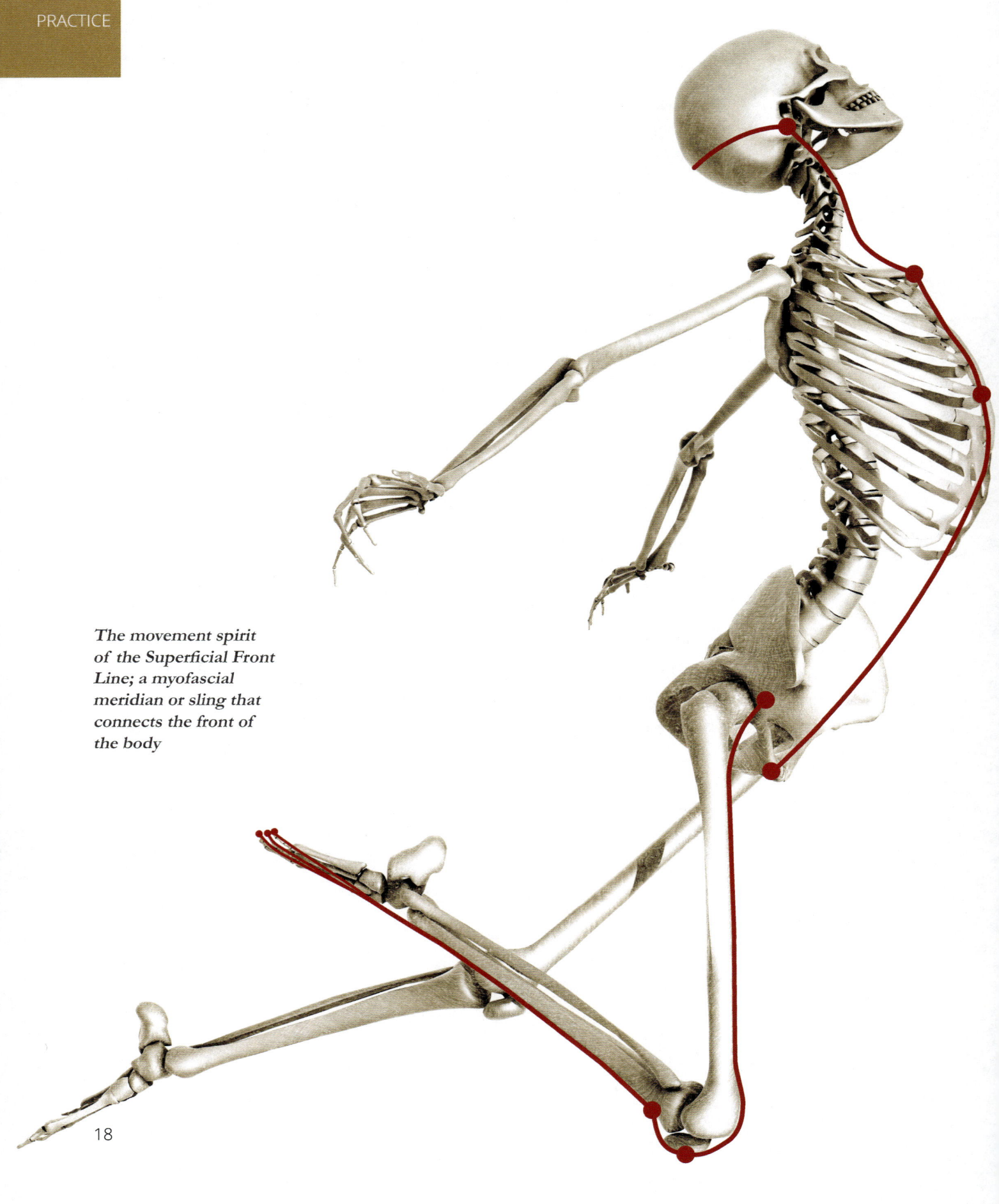

The movement spirit of the Superficial Front Line; a myofascial meridian or sling that connects the front of the body

18

SLINGS PRACTICE

Why Slings? Because it makes you feel and look good from the inside out.

The term 'slings' and the real-life application of the concept require some explanation before wrapping the mind around 'wonderstanding' the intricate nature of fascia and its movement-related qualities.

Slings aka Myofascial Meridians

The term 'slings' refers to internal connections of muscles and fascia. Instead of thinking of individual muscles, visualise long fascial continuities that connect several muscles in series. These continuities form well-toned slings inside your body that connect your feet to your head and your arms to your trunk.

Internal connections of muscles and fascia

If you are familiar with the Anatomy Trains concept by Thomas W. Myers, you can think of a sling as a myofascial meridian; they are equal.

Sling equals myofascial meridian

Deliberately Fascia-Focussed

From the title of this book, you know that Slings is a fascia-focussed practice. You might wonder what that exactly means. It means that we deliberately utilise specific qualities of fascia to improve its overall functionality. For comparison, think of other training modalities. For example, bodybuilding aims to strengthen muscles, and running efficiently improves cardiovascular fitness. Because all movement is myofascial, bodybuilding and running also train certain qualities of fascia. However, the training effects on fascia are typically a by-product, not the primary goal.

Specific qualities of fascia are deliberately utilised

Now that we have established what fascia-focussed means, we need to speak about what kind of fascia-focussed practice Slings is.

Body-Minded, Resource-Oriented, Versatile

Fascia-focussed training can be approached in many ways. It can be more fitness-oriented, performance-driven, and specialised, or it can be like Slings: body-minded, resource-oriented, and versatile. Neither is right or wrong; they are just different approaches to utilising fascia in movement.

PRACTICE

Practicalities

Yoga-style setting

On a practical level, envision a yoga-style setting for a Slings lesson.

- The practice is done barefoot on a mat.
- The movements are performed with awareness and in a deliberate manner.
- The repertoire of exercises is versatile and contrasting, ranging from slow and controlled, to melting motions, to held poses, to rhythmical, dynamic movements.
- Self-massage exercises are incorporated by using relatively soft props, such as textured massage balls and domes.
- Although light weights may be utilised, the practice is generally done with bodyweight only.
- The exercises are adapted to suit individual requirements.

7 Distinguishing Features of the Slings Practice

Functional movement art with integrity

1. Slings is a holistic movement practice that is rooted in Western anatomy.
2. The practice has therapeutic value, yet it is resource-oriented instead of illness-focussed.
3. The inner slings we train are not merely intuitive. They are well-established and clearly defined continuities of muscles and fascia.
4. The practical applications are based on twelve science-informed Fascial Movement Qualities.
5. Fascia is engaged in a versatile manner to improve a broad spectrum of its functions.
6. The practice is multidimensional, moving the whole body in all directions.
7. The lesson structure, sequencing, and exercises are freely adaptable.

6 Good Reasons for Slings

1. **Postural ease:** Adaptable postural balance and the related sense of inner equanimity.
2. **Movement freedom:** Poised and efficient movement as desired or demanded by life.
3. **Movement love:** The inner motivation to move – both in practice and everyday life.
4. **Somatically at home:** The feeling of inner belonging and togetherness.
5. **Meaningful self-awareness:** The conscious experience of feelings and the ability to acquire meaning from what is perceived.
6. **Radiant vitality:** Being and feeling vibrantly alive.

Immediate and long-lasting benefits

These six good reasons for Slings are discussed in more detail at the end of Part 2.

ONTO THE MAT AND INTO THE BODY

Why not get your myofascial slings in motion right now? Take off your shoes, roll out your mat, and grab two soft massage balls to enjoy the diverse selection of complimentary Slings practices that we offer on the art of motion Academy website and YouTube channel.

Complimentary Practice Videos

art-of-motion.com → Cinema → Practice

YouTube → art of motion Academy → Practice – Slings Myofascial Training

FASCIAL SYSTEM IN A NUTSHELL

"Fascia is the fabric of the body, not the vestments covering the corpus, but the warp and weft of the material. Remove all other tissues from their fascial bed and the structure and form of the corpus remains, ghostlike, but clearly defined."

Stephen M Levin and Danièle-Claude Martin

WHAT IS FASCIA?

Collagenous, viscoelastic connective tissue

When viewing fascia holistically, we speak of the fascial system. It includes all of the collagenous, viscoelastic connective tissues in the body, as well as the cells that make them. The fascial system is the continuous, fluid fibre matrix that shapes us, differentiates internal structures, yet connects everything with everything else, from muscles to bones, internal organs, nerves, and blood vessels. It is omnipresent from underneath the skin into the core of the body, and from the feet up into the brain. Even though the fascial system can be divided into individual parts, in our physical reality, it is one whole system where local events have bodywide effects.

4 Fascial Facts

Where is Fascia?	Everywhere
What Does Fascia Connect?	Everything
Can It Be Trained Exclusively?	No
Can It Be Trained Deliberately?	Yes

Should We Forget About the Rest?

Muscle-focussed and neuro-cognitive training is important

Absolutely not! Muscle-focussed and neuro-cognitive training methods emphasising organ health or bone alignment are as important as ever. Deliberate fascial training simply adds a different dimension to movement practices.

Having clarified that, let's start from the beginning by defining the term fascia and exploring what the system encompasses.

ANATOMICAL AND FUNCTIONAL DEFINITIONS OF FASCIA

A globally agreed-upon definition of what the term fascia exactly encompasses is still under discussion. At this point, we have strictly anatomical and functional definitions of fascia that we can choose from depending on context.

Anatomical Definition: Dissectible Parts

From a purely anatomical outlook, fascia is defined as a sheath, or any number of dissectible connective tissues that form beneath the skin to attach, enclose, and separate muscles and other internal organs. Although reductionistic, the anatomical definition of fascia is very useful for understanding specific pathologies, functionality in individual structures, as well as for comparing different research findings. Let's appreciate what we can learn from it!

Functional Definition: Fascial System

The functional or holistic definition of fascia is broader and much more inclusive than the strictly anatomical definition. Instead of an aggregate of dissectible parts, fascia is seen as an integral system that includes all of the body's collagenous connective tissues, as well as the cells that build, maintain, and remodel it. From here onward, we will work with the definition of fascia as an integral system.

Functionally, fascia is one bodywide system

FASCIA

LIVING SYSTEM WITH SPECIALISATIONS

Countless rope-like densities, straps, sheaths, fluid channels, and pockets

The collagen architecture and tissue composition vary greatly within the fascial system. Formed within are rope-like densities, stabilising straps, resilient sheaths, fluid channels, and thousands of pockets, all of them interconnected by sturdy membranes, fluid, and loose layers.

Embedded in this magnificent tissue web are our 650-plus muscles. With its unique qualities and close relationship to the muscular and nervous system, fascia plays a leading role in dynamic stability and movement.

Constantly remodelling according to demand

Fascia is not, as assumed in the past, lifeless packaging and superfluous tissue. As you are reading these words, there is a truly magical spectacle happening inside your body. With the ever-changing fluid dynamics of an inner ocean, your fascia constantly remodels according to demand. There is no inertia or standstill in the fascial system. As you can see, the moment to moment self-regulation of your fascial system is quite an undertaking and accomplishment. By staying in constant contact with trillions of cells, this 'beehive' of active fascia keeps the body in shape and functioning!

Image by Polly Dot from Pixabay.com

FASCIA IN MOTION: UNDERSTANDING IN PROGRESS

"Scientific studies allow only partial understanding of fascia – they can't explain movement in all its aspects."
Carla Stecco at the Fascia Research Congress Washington 2015

Knowledge in development

In the last decade, fascia research has rapidly grown and made considerable leaps in understanding this complex system. Still, it is a relatively young field of study. In science, there are many variables. The knowledge remains incomplete and is in development. This is especially true for our understanding of fascia in the living, moving, feeling body, where it works with all other bodily systems. As of yet, it is impossible to isolate the fascial system for the implementation of controlled studies and challenging to directly assess its nature.

Science translated into practical application is interpretive

It is fair to say that researchers and movers are still in the process of understanding exactly how fascia works in motion. While the search for knowledge continues in labs around the world, those of us intentionally including fascia in our approach have the possibility to translate current scientific information into practical applications as we go. Like many of my colleagues, this is what I have done in the last few years: study, translate, apply, evaluate, connect dots, reconsider, rethink, translate differently, apply differently, re-evaluate. In this perpetual learning process, I have come to understand the value of twelve Fascial Movement Qualities. Based on these qualities, I developed a set of twelve Myofascial Training Techniques designed to achieve defined short-term aims and long-term aspirations.

Is this the only or 'right way' to translate fascia research into practical application? Of course not. The question is not about right or wrong; it is about practical value and whether my translation resonates with you.

Your body manufactures all these materials and many more by mixing together various proportions of the ECM's fibers and glue and altering the chemistry in different ways (Snyder 1975). In bone, the fiber matrix is there—much like leather—but the mucousy ground substance has been systematically replaced with mineral salts. Cartilage has the same leathery substrate, but the glue has been dried into a tough but pliable "plastic" that permeates the fibrous leather. In ligament and tendon, almost all the glue has been squeezed out. In blood and joint fluid, the fiber exists only in a liquid form, until it hits the air, when it forms a scab. This manufactory in your body is fascinating: the dentin in your teeth, your gums, your heart valves, even the clear cornea of your eye—are all formed in this fashion.

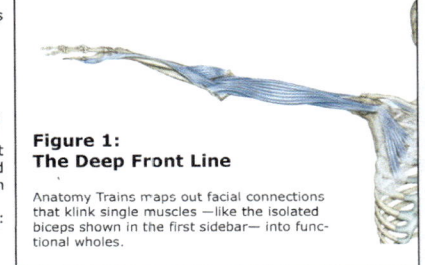

Figure 1: The Deep Front Line

Anatomy Trains maps out facial connections that klink single muscles —like the isolated biceps shown in the first sidebar— into functional wholes.

Remodeling and Tensegrity

Your muscles may determine your shape in the training sense, but connective tissue determines your shape in the overall sense. It holds the bones together, pulling in on them as they press out (like a tensegrity system; see Figure 2).

The ECM is capable of remodeling itself in a variety of ways (Chen et al. 1997). Just as your muscles remodel themselves in response to training, the fascia remodels itself in response to direct signaling from the cells (Langevin et al. 2010); injury (Desmouli`ere, Chapponnier & Gabbiani 2005); long-held mechanical forces (Iatrides et al. 2003); use patterns (including emotional ones); gravity; and certain chemistry within your body (Grinnell & Petroll 2010). The complexities of remodeling are just now being explored in the lab; the details will be revealed over the coming decade.

The idea of **tensegrity** (tension and integrity) and the phenomenon of remodeling are the basis for structural therapy, including yoga and the forms of manual therapy commonly known as Rolfing® or Structural Integration and its deep-tissue relatives, including foam rolling. Change the demand—as we do in bodywork and personal tr[aining]... nds to that new demand. This common theme points to a futu[re]... combine to form a powerful method for

RULE NUMBER ONE
Form follows function, function alters with form.

This means that the architecture of the fascial system shapes with the way we behave and that our behaviours change with the state of the fascial system. In a narrower sense, the organisation of fascial structures such as aponeuroses, retinacula, ligaments, tendons, muscle fascia, and loose fascia forms in accordance with the way we move, while our movements alter with the organisation of fascial structures.

You and I, as well as everyone else, have the same fascial components. However, the way our fascial system is structurally shaped is unique. Imagine you and three of your friends are given the opportunity to design a house. Each of you is given the same composites in a different ratio. By the end of the exercise all of you have designed a house with unique architectural features that represent your tastes, skills, needs, wants, creativity, and more. Based on their architectural features, the houses enable or require a different lifestyle. Like a house, the fascial system needs maintaining to avoid disintegrating and it can be remodelled to facilitate an alternative kind of living.

THE COMPONENTS OF FASCIA

The fascial system is made up of fibres, ground substance, and water. In our inclusive view, we also consider cells and sensory receptors to be an integral part of fascia.

Fibres
Fascia contains collagen and elastin fibres. Collagen is the most prevalent fibre type and, together with water, is the most abundant component of fascia.

Ground Substance
The ground substance is a gel-like material that consists of molecules that absorb water, similar to the absorptive property of sponge.

Water
Water is an essential ingredient of fascia. Most of the water in fascia is bound, like the fluid absorbed by a wet sponge.

Cells
The fascial system also contains the cells that make it, mend it, break it down, lubricate it, and facilitate communication within.
The cells in focus are the contractile fibroblasts, their specialised siblings, the myofibroblasts, and the lubricating fasciacytes.
Not explicitly discussed are the telocytes, which are believed to aid intercellular communication and cell repair.

Sensory Receptors
Fascia is highly innervated. The sensory receptors in focus are the Golgi, Pacini, Ruffini, and interstitials.

COLLAGEN FIBRES

Collagen fibres are whitish. Therefore collagen-rich tissue is correspondingly fair in colour. After water, collagen is the second-largest component of fascia.

Structure, stability, tensile strength

Collagen gives fascia its structure, stability, and tensile strength. The more collagenous the tissue is, the firmer it is. As a force transmitter it is more stable, stronger, and efficient than loose, watery fascia. To withstand strain and to distribute forces most effectively, the collagenous architecture forms and remodels according to use or how it is loaded.

- Collagen fibres that are repeatedly loaded in a predominantly linear fashion orient themselves according to the relatively linear line of force. The resulting unidirectional collagen architecture gives the tissue great stability, tensile strength, and resilience in a certain direction, as seen in tendons and ligaments.

- When fascia is regularly tensioned from different vectors, the collagenous architecture is much more multidirectional, as seen in the double lattice arrangement of muscle fascia.

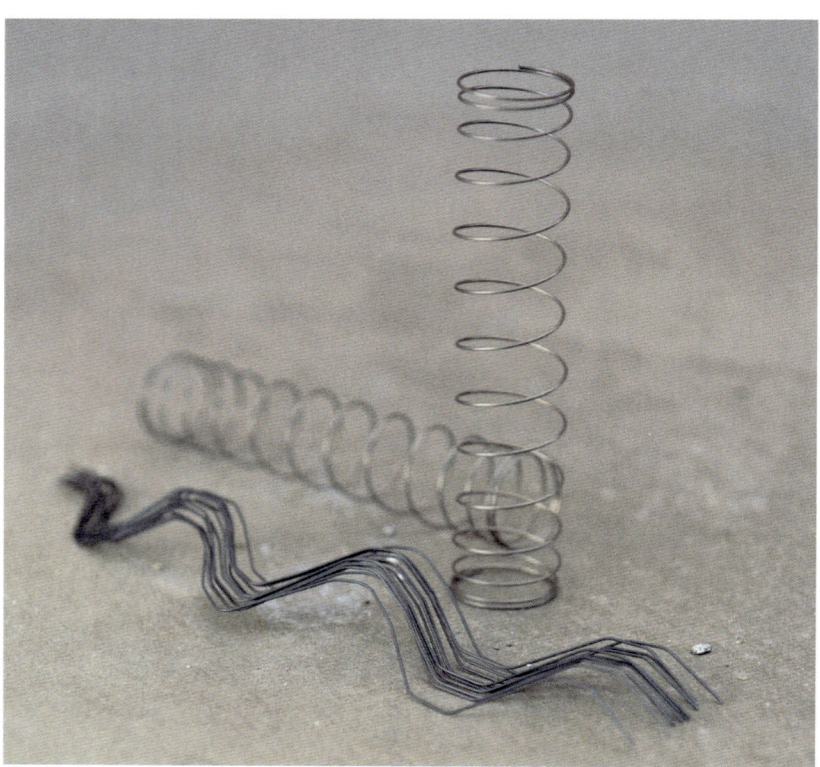

Resilient Architecture

Highly collagenous fascia has great potential for kinetic storage capacity. Meaning, when the collagenous architecture is resilient the tissue has elasticity. Two key features of resilient collagen are its spiralling organisation and crimp formation.

Spiralling:	Collagen essentially consists of three long protein chains that form a triple helix. These spiralling molecules form a fibril, which together with other fibrils assemble to become a collagen fibre. When the fibre is lengthened and therefore tensioned, spirals intertwine, which increases the fibre's steel-like strength.	*Steel-like strength*
Crimp:	Collagen in fascial structures that are elastically used have crimp, which is an undulating fibre orientation when the tissue is not tensioned. With an increase in crimp, fascial elasticity increases.	*Tissue elasticity*

Functional Qualities

Collagen density and architectural patterns are determining factors in tissue stability and elasticity.

Dynamic stability: Well-organised fascia provides dynamic – or adaptable – stability for the body. The higher the collagen content is in the tissue, the firmer and more stable it is.

Elasticity: Crimp formation and a resilient collagen architecture are two key features of elastic fascia. Together, they significantly enhance the kinetic storage capacity of the tissue and therefore facilitate buoyancy and spring in rhythmical and dynamic motions.
You can visualise collagen-rich fascia like a tough elastic band that can stretch a bit and then recoil at high speed, affording great efficiency to movement.

FASCIA

ELASTIN FIBRES

Yellowish in appearance, elastin fibres are very stretchy, hence their name. This fibre type is abundant in fascia and other tissues where elasticity is of major importance; for example, in cartilage, blood vessels, the lungs, and the skin.

Organisation and Functional Qualities

The orientation of elastin fibres is multidirectional. They branch out and have many connections to each other, forming criss-crossing networks.

Multidirectional, very stretchy fibres

Elastin fibres can be stretched to about 200% of their resting length. To be clear, when elastin is stretched, the fibres (or more accurately, the molecules within the fibres) don't actually stretch. Instead, they straighten in the direction of pull until they are fully tensioned. When the load is released, they recoil to resume their original shape.

Multidirectional elastic adaptability

Elasticity: Like collagen, elastin fibres can store and release kinetic energy.
Picture a stretchy rubber band that can extend to double its length and then recoil.

GROUND SUBSTANCE

Ground substance (GS) forms the non-fibrous, fluid part of the fascial system. It is a gel-like, transparent, and colourless material within which fibres are woven, and cells are embedded.

Gel-like and water-absorbing

Composition and Organisation

The ground substance is made up of large macromolecules called proteoglycans (PGs), which contain molecules called glycosoaminoglycans (GAGs). A family of seven distinct glycosoaminoglycans is recognised: hyaluronan, chondroitin-4-sulfate, chondroitin-6-sulfate, dermatan sulfate, keratin sulfate, heparin sulfate, and heparin. To explain their organisation, proteoglycans are oftentimes compared to a bottlebrush, though they also resemble a conifer branch.

Image by Mabel Amber from Pixabay.com

Binding water like a sponge

Functional Qualities

Unlike bottlebrush bristles or conifer twigs, proteoglycans bind a large portion of water like a sponge that absorbs fluid. In that sense, they determine the water content in fascia and the degree of viscosity in the tissue. Viscosity refers to the resistance of a fluid to flowing (how quickly it changes shape in response to external forces) and its stickiness (how strongly it adheres to adjacent structures). For example, syrup is more viscous than water. The fluidity of the ground substance acts as a lubricant that enables collagen fibres to slide against each other without friction until the interfibrillar cross-links are tensioned. In this way, ground substance facilitates movement ease by absorbing forces that affect the tissue, and it protects the collagen fibre network from excessive strain.

Facilitating movement ease

Ground substance is also responsible for the support and nutrition of cells while providing some mechanical support by joining cells together. Lastly, its viscoelasticity allows the tissue to return to its original shape after deformation.

Viscosity:	GS is a gel that can be fluid or dense.
Elasticity:	GS is resilient.
Glide:	GS enables frictionless glide between fibres.
Shock absorption:	GS absorbs compressive forces.
Protection:	GS protects the collagen network from excessive strain.
Stability:	GS provides a degree of intercellular stability.
Cell nourishment:	GS facilitates intercellular exchange.

Image by Robert Fotograf from Pixabay.com

WATER

A major component of fascia, water is distributed throughout the system. The fluid in fascia is commonly called interstitial fluid or fascial water. A large quantity of fascial water is not free-flowing, instead, it is bound by the molecules that make up the ground substance.

Bound and free-flowing

Image by mac231 from Pixabay.com

Functional Qualities

Well-hydrated fascia is vital for pretty much everything, from cellular health to movement functionality.

Supporting vitality and functionality all around

Health:	Water is essential for all physical functions.
Transportation:	Water is an important transportation medium for removing waste products and distributing nutrients.
Temperature:	Water assists in the regulation of body temperature.
Communication:	Water is a communication system that circulates messenger molecules.
Glide:	Water adds fluidity to ground substance, which facilitates movement ease.

FASCIACYTES

Traditionally, secretion of the viscous ground substance has been ascribed to the most common cell family in fascia, the fibroblast cells. It would stand to reason to discuss them now. However, in 2010, Carla Stecco and her team discovered a new cell family whose primary role is to support the fluidity of fascia. Let's transition from the topic of fascial water directly to those cells, the fasciacytes.

Functional Quality

Devoted to lubrication

Research findings suggest that fasciacytes are devoted to the production of hyaluronan, a key ingredient of ground substance. This particular type of glycosaminoglycan binds water and forms a lubricating gel that facilitates smooth sliding motions between adjacent myofascial structures. Fasciacytes populate the loose fascia amidst deep fascial layers, epimysia, and the sliding fascia inside of muscles.

Facilitating myofascial glide

Glide: Fasciacytes facilitate glide between and within muscles, which supports movement ease and functionality.

Martina, Alexa, and Karin feeling in the flow in the Indian Ocean at South Beach in Western Australia

FIBROBLASTS AND MYOFIBROBLASTS

FIBROBLAST CELLS

The fibroblasts are the most abundant cell-line in fascia.

Functional Qualities

By secreting collagen and elastin proteins, fibroblast cells constantly remodel the insoluble, fibrous network of the fascial system. For these interconnected cells to stay healthy, they need regular tensioning and, therefore, movement. In turn, when there is an injury, fibroblast cells generate traction to the tissue to support wound healing.

Devoted to remodelling

Structure: Fibroblast cells remodel the fibrous architecture of fascia.

Facilitating structural changes

Stability: Fibroblasts assure a healthy degree of fascial firmness, tautness, and adaptability.

Wound healing: When tissue is damaged, fibroblasts aid wound healing.

Karin and Martina building, removing, and mending things at Fremantle Harbour in Western Australia

MYOFIBROBLAST CELLS

Remodelling,
tone regulation,
reinforcement of fascia

The contractile myofibroblasts are activated fibroblast cells whose primary function appears to be the remodelling of the fibrous network and fascial tone regulation. The latter is often associated with tissue reinforcement. While the exact mechanisms are not yet fully understood, it is known that fibroblasts can convert into differentiated myofibroblasts. Along with other factors, it is mechanical changes that induce the differentiation from fibroblast to myofibroblast.

For a long time, myofibroblasts were exclusively associated with wound healing and pathological contractures. It was in the early nineties that researchers discovered cells in deep fascia that contracted when the tissue was isometrically tensioned. Just a few years later, smooth, muscle-like cells were also reported in deep fascia. Within their vicinity, a large number of sympathetic nerve tissue and sensory nerve endings were found. This led to conclusions regarding the connection between myofibroblasts, the autonomic nervous system, and fascial tone regulation independent of tonus in the muscles.

Involuntary Contractility

Involuntary, very slow
contractions

Myofibroblast cells can contract in a similar way to smooth muscle, though at a very slow speed.

Involuntary: Myofibroblast contractions are involuntary, therefore beyond your conscious control.

Contractile: The contractile capacity of a myofibroblast is greater than that of a regular fibroblast cell.

Very slow: Unlike skeletal muscle cells, myofibroblasts contract very slowly, within a timeframe of minutes.

Although myofibroblast cells only induce small tissue contractions, over time, the cumulative effect can lead to significant tissue stiffness. Naturally, the effect goes the other way around too. A decreased density of active myofibroblasts lowers fascial tone.

Functional Qualities

Wound healing: Myofibroblasts have important wound healing functions.

Tone regulation: They regulate the basal tone of fascia.

Reinforcement: Where functionally required, myofibroblasts reinforce tone in healthy tissue.

Karin and Martina feeling empowered by fascia in Fremantle in Western Australia

SENSORY RECEPTORS

"Any intervention on the fascia is also an intervention on the autonomic nervous system."
Robert Schleip

The fascial system and nervous system are intimately linked

Fascia is highly innervated. Besides its many other roles, it is an influential perceptual system that functions as an inner sixth sense. It facilitates reflexive and wilful muscle coordination, fascial tone regulation, the conscious perception of feelings, as well as health-oriented behaviour adaptations. This sensory function of fascia is instrumental for experiencing the body and recognising emotional states, interpreting what is felt, and adapting in a health-oriented manner.

As of the writing of this book, it is estimated that the number of nerve endings in the fascial system is about 250,000,000. If this number has no meaning for you, just know that it is a lot! There is still much to be learned about locations, density variations, and roles of the vast amount of sensory material embedded in our fascia. However, the intimate relationship between the fascial system and the nervous system (somatosensory and autonomic) is undeniable. For this reason, nerve elements are taken into account in our holistic perspective.

While our attention is on sensory receptors that contribute to proprioception (movement coordination) and interoception (feeling awareness), the presence of sympathetic nerve fibres in fascia is worth noting. Sympathetic nerves, also called 'C fibres' or 'small fibre' nerves, contribute to temperature perception and are of special interest in the treatment of pain that is described as burning, achy, tingling, or numbing in character.

Sensory Receptors

The following four mechanoreceptors are found in fascia without being exclusive to it.

Golgi: The Golgi apparatus respond to slow stretch signals, impacting muscle tone.

Pacini: The Pacinian corpuscles respond to rapid pressure changes, as well as vibration.

Ruffini: The Ruffini corpuscles are slow adapting receptors that respond to stretch.

Interstitial: The interstitial receptors are abundant free nerve endings. They respond to a broad variety of stimuli, including dynamic or sustained pressure changes with intensities ranging from firm to feather-light, sensual touch.

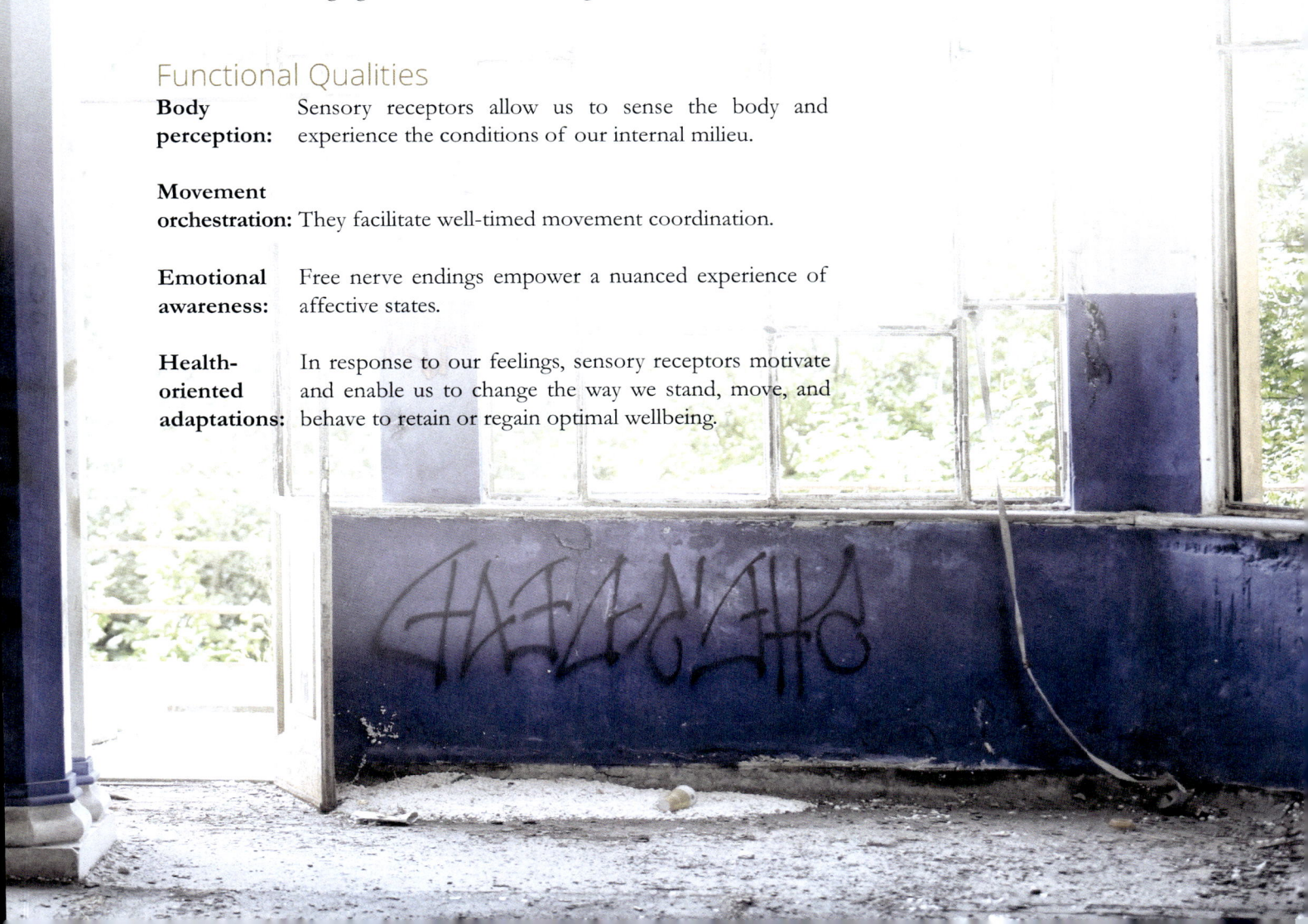

Functional Qualities

Body perception: Sensory receptors allow us to sense the body and experience the conditions of our internal milieu.

Movement orchestration: They facilitate well-timed movement coordination.

Emotional awareness: Free nerve endings empower a nuanced experience of affective states.

Health-oriented adaptations: In response to our feelings, sensory receptors motivate and enable us to change the way we stand, move, and behave to retain or regain optimal wellbeing.

SOMATIC FASCIA DIFFERENTIATED

As previously stated, the fascial system encompasses all collagenous, viscoelastic connective tissues of the body. This includes the fascia associated with the visceral and nervous systems. In terms of study, you are looking – without exaggeration – at a gigantic scope. For practical purposes, our focus is 'only' on the somatic fascia. In other words, the fascia that interacts directly with skeletal muscles and bones.

So, when referring to the fascial system in this book, the reference is specifically pointing to the somatic fascia. At the same time, there is an acknowledgement that changes in this part of the system have an effect on the internal organs, the brain, and the countless nerve branches that permeate fascia from the inside out. And vice versa.

Nomenclature in Development

"Taxonomies should be working for us, our purposes and applications, rather than us being enslaved to them."
Gil Hedley

There is no universal agreement on anatomical terms relating to fascia yet. Sure, there is a common understanding of what the plantar fascia and thoracolumbar fascia are. However, if we take it just one step further, expert opinions about nomenclature diverge already. The same goes for the defined naming of fascial specialisations or what exactly is covered by a term. This makes for an exciting time of potential pronouncement. Meanwhile, though, it also leads to a great deal of confusion and regular misunderstandings between professions and professionals. The subject matter being complex as it is, language should assist effective communication rather than hinder it by presenting an added challenge. To convey the information clearly in this book, we need to establish a common fascia-related language. Therefore, I have linguistically subdivided the fascial system into four specialised groups, each with a representative name that is also commonly used by other professionals. Within each group are individual fascial structures that are labelled according to standard or newly developed anatomical nomenclature. In case multiple anatomical names are available, the most common or most descriptive name was chosen.

Defining Fascia

In your physical reality, you have one fascial system. Within the fascial system, specialisations form, which create the unique structural architecture of your body. Fascial architecture forms and remodels according to how it is used to meet the varying demands placed on the body. Fascial specialisations are expressed as fine membranes, fluid layers, resilient sheaths and straps, sturdy ropes, and more. A few centuries ago, these specialisations were labelled by early anatomists, many of them studying and defining anatomy the way we know it today. Nowadays we agree that fascial tissue connecting a muscle with a bone is called a tendon and that the tough strap that runs across the heel is the Achilles tendon. We also agree that the iliotibial band is a highly collagenous connective tissue band within the fascia lata of the thigh, which is continuous with the crural fascia below the knee, and the thoracolumbar fascia above the hips. Yet where one exactly ends and the other begins remains a matter of interpretation.

Artificial as these differentiations are from a somatic viewpoint, they are incredibly practical for a shared understanding of what they are. Therefore, we will call a tendon a tendon and use anatomical names like Achilles tendon, all the while acknowledging that they are specialisations within one continuous fascial system. Broader terms like 'element', 'layer', 'membrane' or 'envelope' are also used in this book because they are descriptive.

Understanding Characteristics

To understand the differentiation and functional aspects of fascial specialisations and what we call fascial types, it is helpful to consider their location, makeup, and characteristics.

Location:	Where is the fascial structure?
Architecture:	What is the fluid-fibre ratio and what are the collagen patterns?
Functions:	What are the roles in the body and the functions for movement?
Relationships:	What are the neighbouring and distant connections?

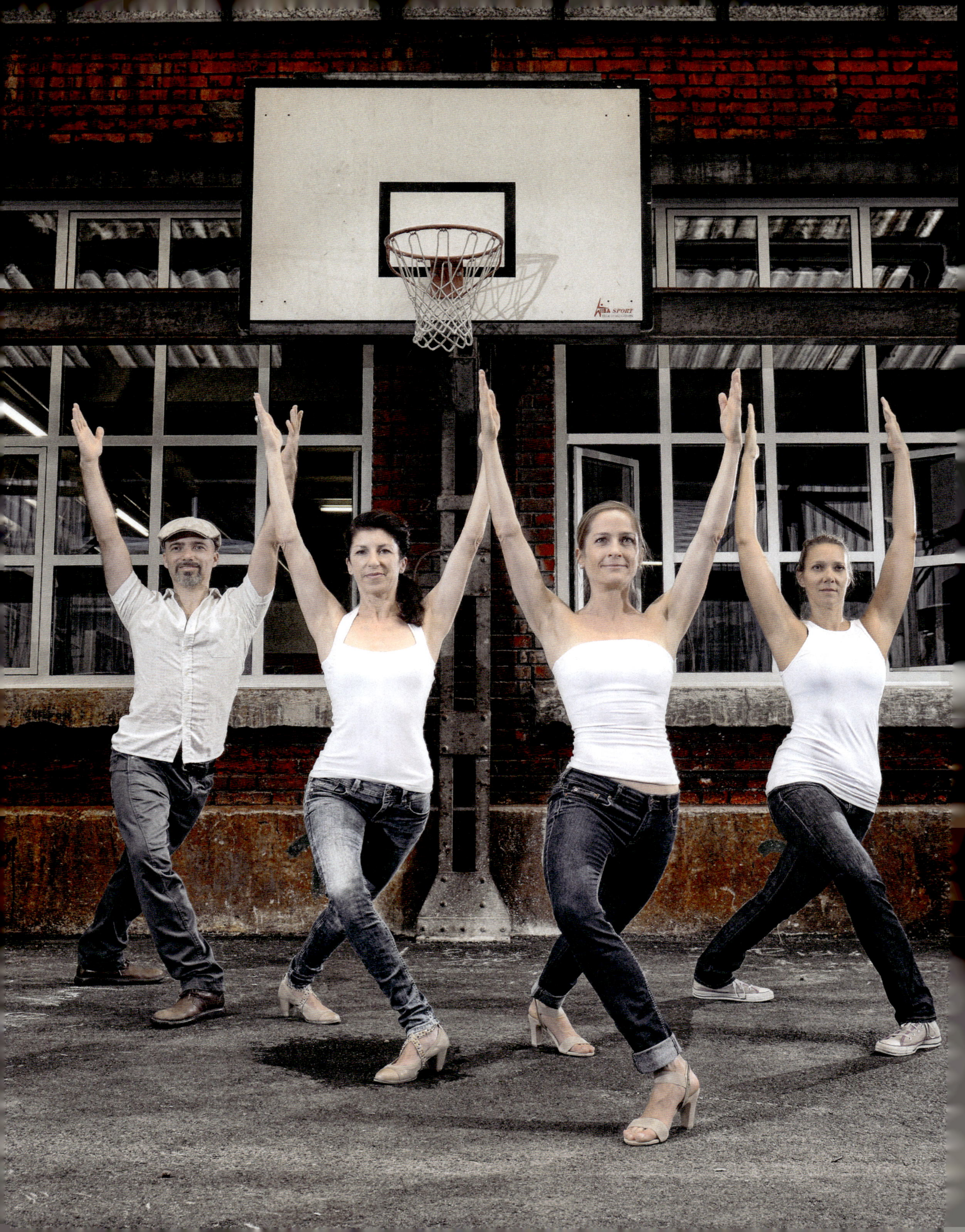

4 DEFINED FASCIAL TYPES

While fascial specialisations express as a local adaptation, for example, as a ligament, tendon, or the epimysium of a muscle, what we call 'fascial types' has a much broader meaning. Fascial types refer to bodywide specialisations in which local specialisations form. For example, a tendon is a specialisation within the 'deep fascia', which is one of four defined tissue types in the fascial system.

Superficial Fascia
The superficial fascia is a multilayered structure beneath the skin that includes the adjoining adipose tissues.

Loose Fascia
Loose fascia is a multidirectional, watery network that is ubiquitous in the body. It provides sliding layers and flexibly links adjacent myofascial structures.

Deep Fascia
Deep fascia is used as an umbrella term for highly collagenous fascia that has a well-structured fibre architecture.

Muscle Fascia
Muscle fascia is 'strictly' speaking under the umbrella of deep fascia. However, for functional purposes, in Slings it is given its own category and is therefore used as another umbrella term. It describes the fascia in which the muscle fibres are embedded, namely the epimysium, perimysium, and endomysium.

Reference
In Part 4 of this book, you will find a more detailed description of the four fascial types, including references to the Fascial Movement Qualities and Myofascial Training Techniques in focus.

SKIN TO BONE LAYERING MODEL

To have a clearer mental image of structural relationships, let's have a look at the layering of tissues. We will go from the outside in, from skin all the way down to the bone. Keep in mind that, like every diagram, the following is just a simplified model of our complex, multidimensional human anatomy. Also, the listing of anatomical structures is limited to what is immediately relevant to the explicit movement context of this book.

The skin to bone layering model is intended to help you establish a clearer picture of structural relationships and with it, a refined understanding of exercise purposes. For example, when envisioning the consecutive gliding of myofascial layers in domino motions or the targeted structures in self-massages.

Note: Please recall what is said in the section entitled nomenclature in development. What I have done here, in the fascial layering model, is look at current terminology from the forerunners in fascia research and education. Next, I have decided on language that is current, yet also useful for application in the Slings approach. The truth is, terms that are used today (in June 2020 as the first edition of the book is being published) may be archaic in the months and years ahead. Approaching this organic development with curiosity and a pinch of excitement calms the mind and allows focussing on what I believe matters more than wording, which is meaning.

Skin

The skin has two distinct layers: the epidermis and dermis.

Epidermis: The epidermis is the outer layer of the skin. Made from epithelial cells, it is not considered a fascial structure.

This superficial skin layer forms a semipermeable boundary between your inner body and the outer world. With its exteroceptive quality, it enables you to feel external stimuli, including touch, like the warmth of the sun or the hand of another person on your skin.

Dermis: The dermis is the inner layer of the skin. It is made of fibrous collagenous connective tissue and can, therefore, be considered part of the fascial system. In this view, it is the outermost layer of fascia in your body.

While the epidermis is highly exteroceptive, the dermis is highly interoceptive. Meaning, the outer layer of the skin receives information from the outer world, and the inner layer of the skin receives information from the inner world. Both pass on the information to the autonomic nervous system where it is integrated and interpreted, for example, into feelings of warmth and touch.

From the superficial fascia, small fibrous strands reach up into the dermis, linking them in a somewhat flexible manner.

Superficial Fascia

The term superficial fascia includes the adipose and tough fibrous layers beneath the skin.

Superficial adipose: The superficial adipose tissue is immediately beneath the skin. It is a pliable layer that contains fibres (retinaculum cutis superficialis) and adipocytes, or fat cells. The fibres provide an adaptable connection between the skin and the fibrous layer of the superficial fascia.

Fibrous layer: This multilayered membrane between the superficial and deep adipose tissues is highly collagenous and, therefore, robust.

Deep adipose: The deep adipose tissue is once again a pliable layer that contains fibres (retinaculum cutis profundus) and fat cells, as well as lymphatic tissue. The fibres provide a flexible link between the fibrous layer of the superficial fascia and the deep fascia below.

The superficial fascia, varying substantially in thickness, forms an adaptable bodysuit between the skin and the deep fascia.

Loose Fascia

Areolar tissue: The loose fascia connects the superficial fascia with the underlying deep fascia. The degree of adaptability or adhesiveness varies in different parts of the body and among individuals.

Deep fascia

The deep fascia beneath the superficial fascia covers deeper myofascial layers and forms compartments as well as enhancements such as retinacula from within.

Aponeurotic fascia: In this model, the term aponeurotic fascia points to the most superficial layer of deep fascia. It is made of well-defined fibrous sheaths that cover and keep muscle groups in place. Additionally, aponeurotic fascia can form the aponeuroses of broad muscles.

You can envision the aponeurotic fascia as a thin, sturdy bodysuit that also forms compartments.

Loose Fascia

Areolar tissue: The loose fascia connects the aponeurotic fascia with the epimysium of the underlying muscles in an adaptable manner.

Muscle Fascia

Epimysium: The epimysium is the outermost layer of the muscle fascia. It contains what we know as the 'muscle'.

Perimysium: The intermuscular perimysium envelops the muscle fibre bundles and allows glide between the fascicles within the muscle.

Endomysium: The endomysium is a honeycomb-like structure that encases each muscle fibre within the muscle fibre bundles.

Muscle Fibres

Muscle fibres: Embedded in the muscle fascia are the muscle fibres: the actin and myosin filaments.

Deep Fascia

The following deep fascial structures are layered, yet also connected in series with the muscle fascia and with one another. Note that deep fascial structures arranged in parallel are flexibly linked with loose fascia.

Tendon, Aponeurosis: A tendon and an aponeurosis are fascial specialisations that connect a muscle with the periosteum of the adjacent bone(s). You can view a tendon as a strap-like structure and an aponeurosis as a sheath-like accumulation of deep fascia.

Ligament: A ligament is a fascial specialisation that links two adjacent bones. Most ligaments in the body are serially connected with the adjoining tendon.

Joint capsule: The outermost layer of the joint capsule is fibrous, therefore part of the same collagenous net.

Periosteum

Periosteum: The periosteum is a double-layered, fibrous membrane that connects the tendinous and ligamentous fascia with the bone itself.

Bone

Bone: With our arrival at the bone, we have reached the inner portion of the skin-fascia-muscle-bone layering map.

FASCIA

LAYER MODEL

The structures considered to be fascia are labelled in black font. The structures that are not considered, or specifically named in teaching as fascial, are shown in grey font.

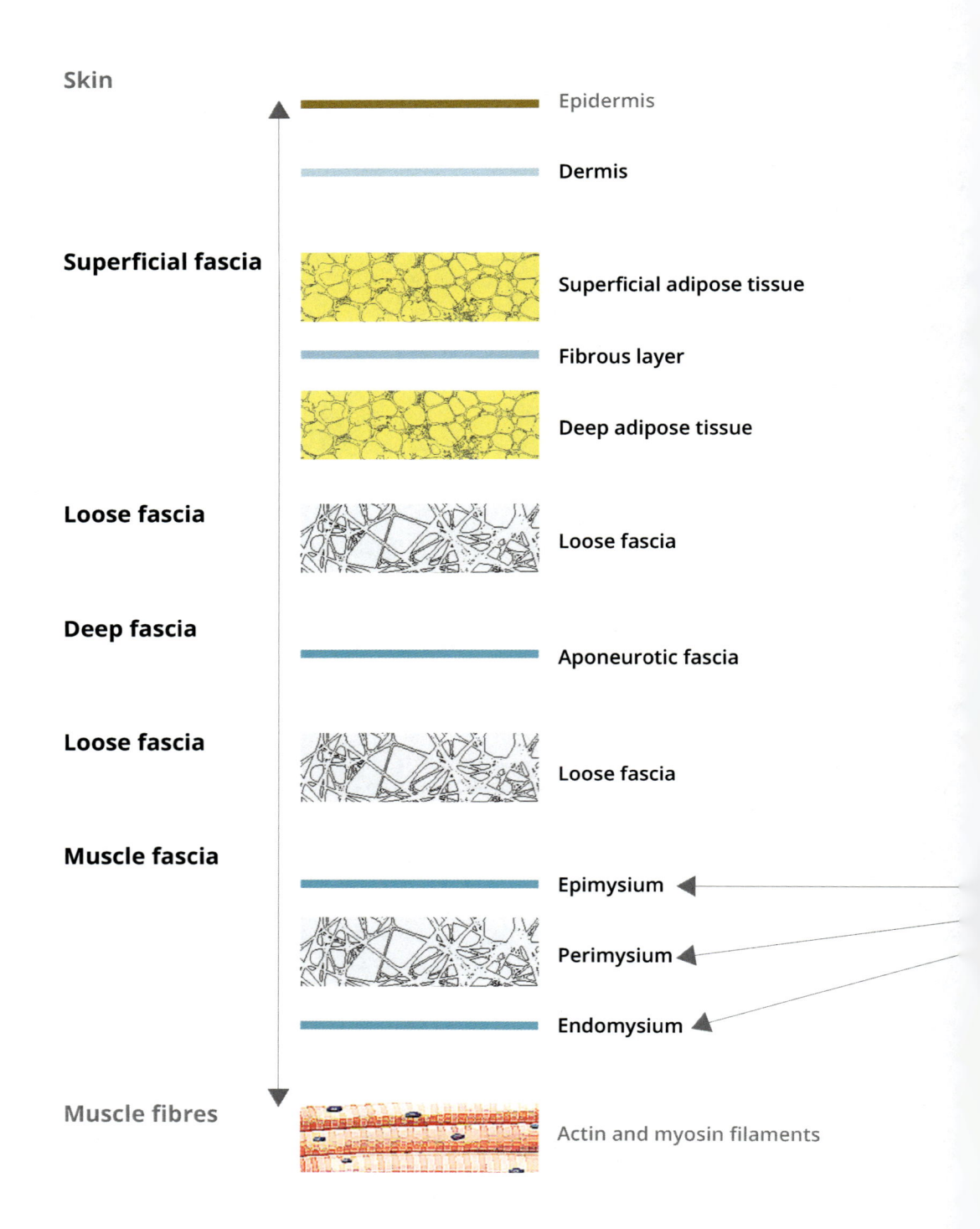

Skin	Epidermis
	Dermis
Superficial fascia	Superficial adipose tissue
	Fibrous layer
	Deep adipose tissue
Loose fascia	Loose fascia
Deep fascia	Aponeurotic fascia
Loose fascia	Loose fascia
Muscle fascia	Epimysium
	Perimysium
	Endomysium
Muscle fibres	Actin and myosin filaments

BODY LAYERS

To understand the integral nature of fascia and related movement applications, it is also helpful to visualise the fascial specialisations as body layers we wear like well-fitting, comfortable clothing.

Skin: A stretchy, sensitive ninja suit.

Superficial fascia: An adaptable, sensual winter neoprene wetsuit.

Aponeurotic fascia: A snug, well-toned summer wetsuit.

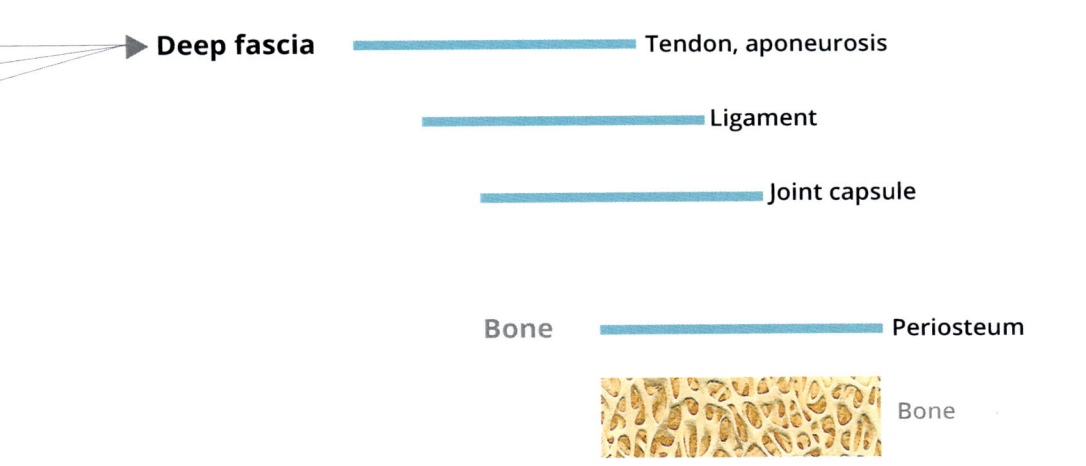

Deep fascia — Tendon, aponeurosis

Ligament

Joint capsule

Bone — Periosteum

Bone

PART 1
SLINGS MYOFASCIAL TRAINING CONCEPT

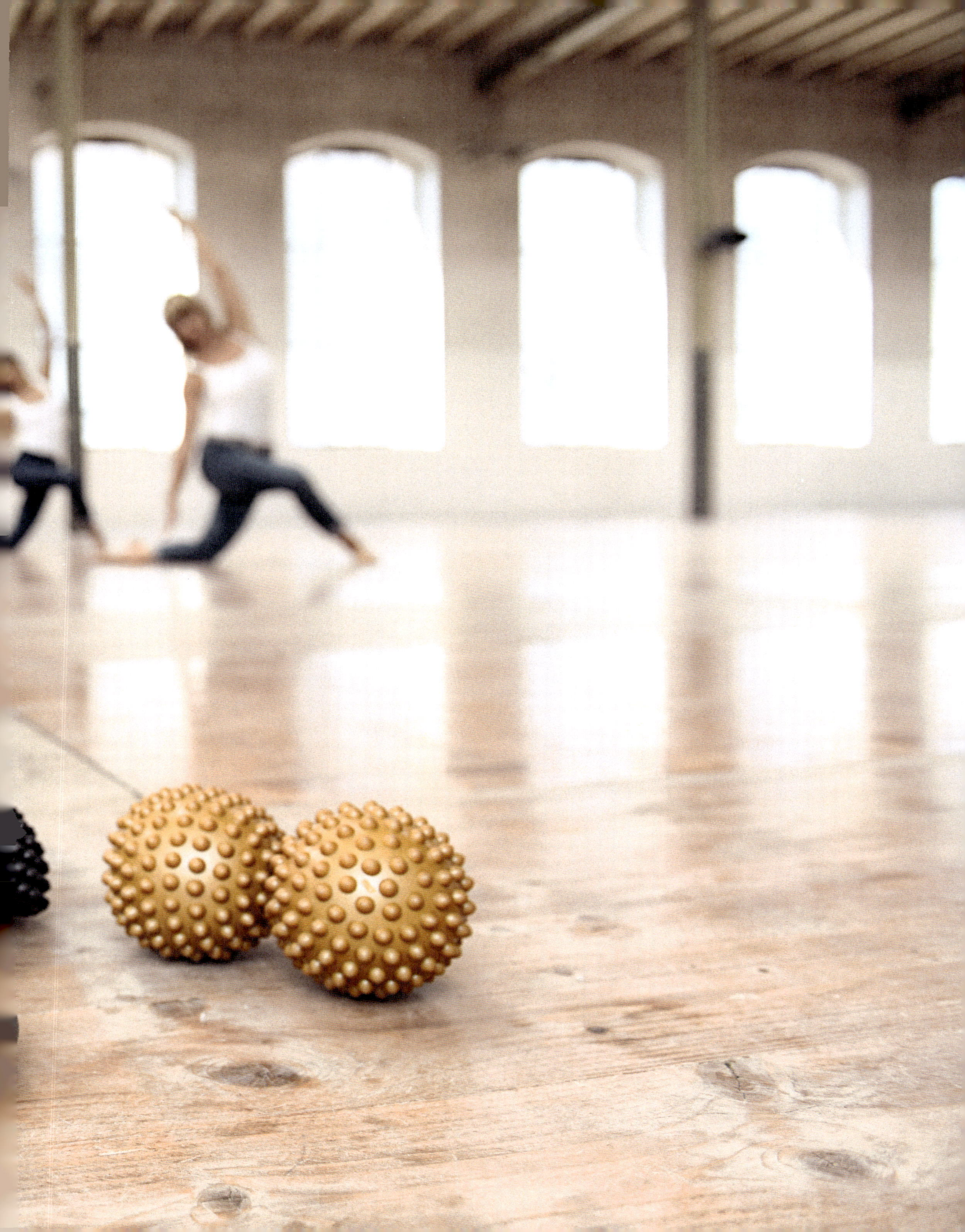

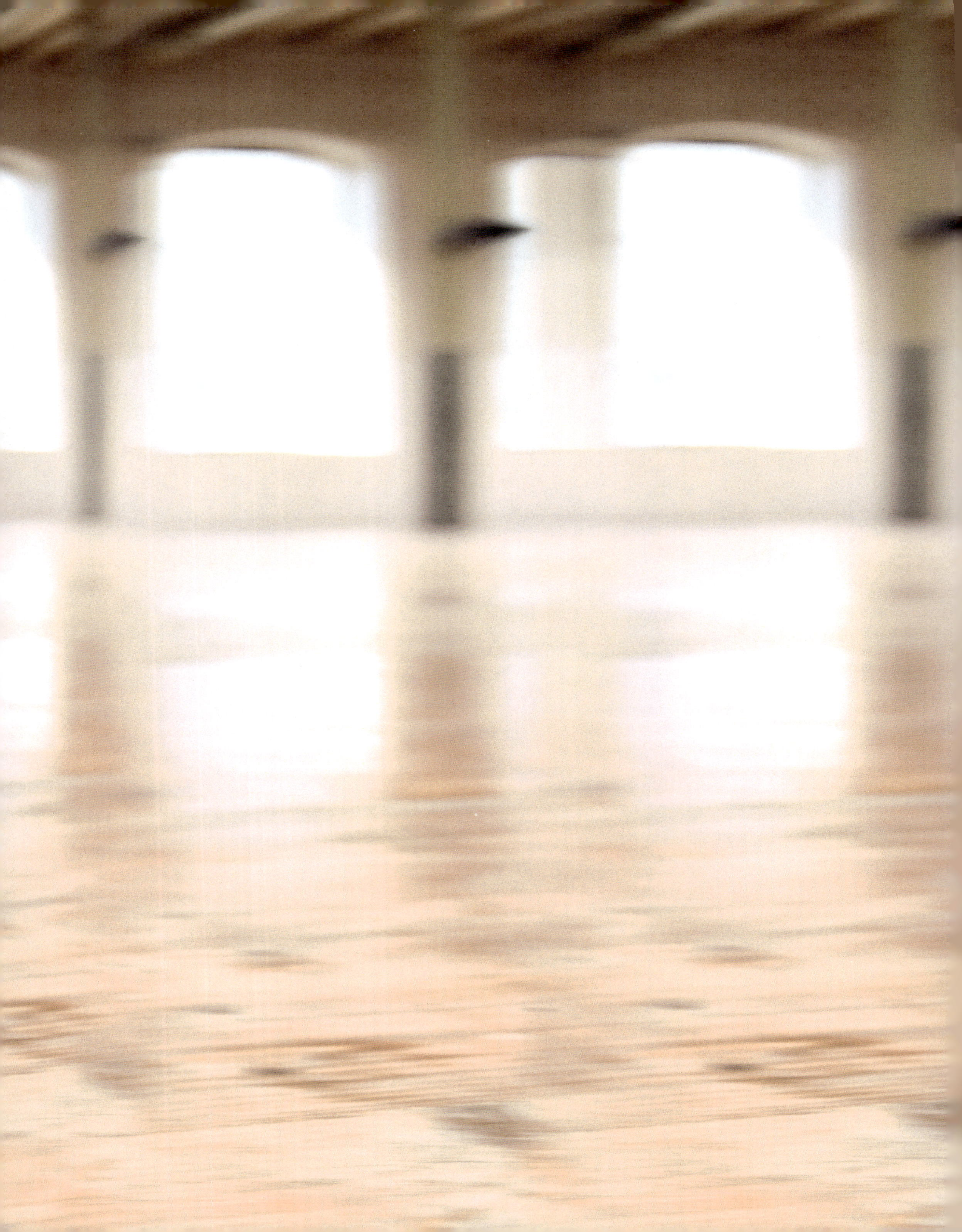

CHAPTERS OF PART 1

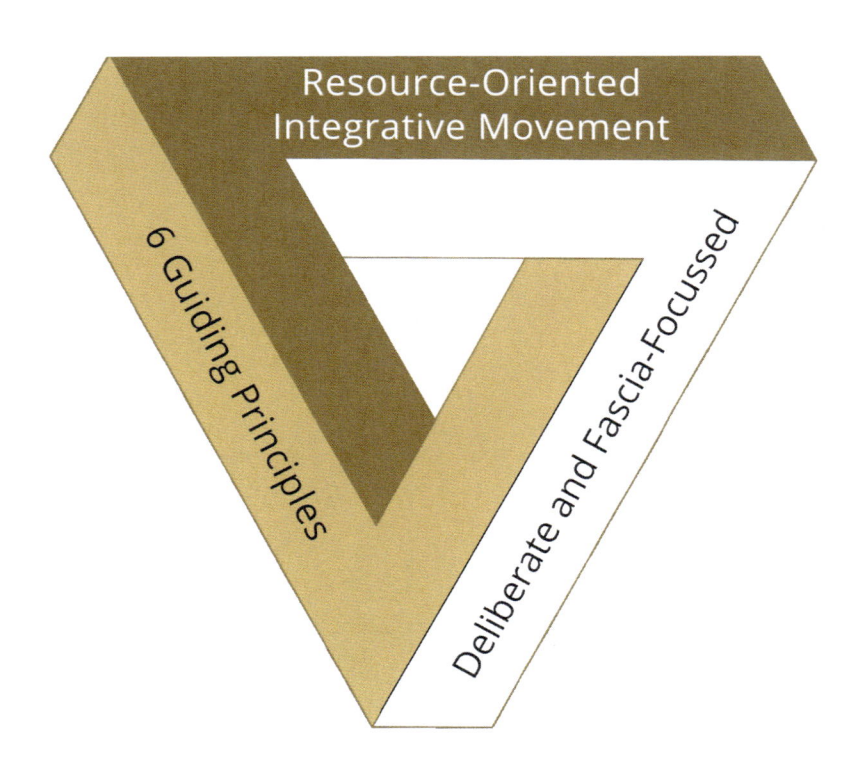

Resource-Oriented Integrative Movement

6 Guiding Principles

Deliberate and Fascia-Focussed

PART 1

RESOURCE-ORIENTED INTEGRATIVE MOVEMENT

Health-centred, uniting, inclusive

Slings Myofascial Training is what we call Resource-Oriented Integrative Movement. Said the other way around, Resource-Oriented Integrative Movement is the umbrella term for Slings Myofascial Training. Before offering you the definition, I would like to state one thing: I didn't wake up one glorious morning with this four-letter term lit up in my mind. Not at all. The phrase was coined only recently, through extensive conversations with Kiki and Heidi, two of my friends and the editors of this book. Our definition deeply resonates with me. Not only because it depicts the essence of the Slings approach faithfully; it also is the result of an eight-year distillation process fuelled by experience and reflection.

Resource-oriented: To utilise somatic sources that enhance the quality of life by focussing on existing abilities, which facilitate meeting and handling physical, emotional, and mental demands.

Integrative: To coordinate and unite the body's systems, emotions, and thoughts to function as a dynamically balanced whole.

Movement: Deliberate, fascia-focussed practice that actively promotes physical vitality, emotional clarity, and mental health.

Slings Concept

Rain or shine, Karin strengthens her resources – this day in front of the Paul Klee museum in Berne in Switzerland

Extended Meaning

Besides the above definition, it is useful to know that 'integrative' in this context refers to more than movement itself. While the movement sequences are integrative, the term also indicates the integration of the twelve Fascial Movement Qualities and the twelve Myofascial Training Aims. The meaning even extends beyond the Slings concept and points to the synthesis with other movement methods and bodywork modalities.

PART 1

DELIBERATE AND FASCIA-FOCUSSED TRAINING

There is nothing new under the sun, except for how the story is told.

Fascia training is ancient, yet nowadays clarity of intent is unprecedented

When throwing a discus, the Greeks utilised the elasticity of their myofascial meridians superbly and the ancient yogis deeply connected to the sensory qualities of their fascia when practising asanas. As said in the Book of Ecclesiastes, "there is nothing new under the sun". This is equally true when it comes to training fascia.

What is new is the intention or focus with which fascia is trained nowadays. With increasing scientific insights into fascial composition, architecture, and their workings, we are able to train fascia more consciously and deliberately. When we understand why the ancient Greeks were such superb sportsmen and why the yogic originators came up with asanas that condition so much more than muscles, we can reap the benefits for our own practices.

Ryan under the sun on the Great Wall of China near Beijing, China

Understanding 'Why' to Determine 'How'

All movement is myofascial, but not all movement modalities train fascia in the same deliberate way or with the same intention. The question is not whether fascia contributes to or is engaged in movement. The question is: How do you employ its properties intentionally to improve fascial functionality purposefully?

Take muscle-focussed training, for example. To train muscles in a specific manner, you need to define what you want to achieve: everyday strength, endurance, power, speed, stability, flexibility, relaxation. Then you need to understand the properties and workings of muscles to train them effectively. I believe the same is true for deliberate, fascia-focussed training; hence I defined twelve Fascial Movement Qualities that are based on established fascial properties.

To be clear, I don't believe that there is such a thing as 'the' fascia training, or the best, or the most effective way to train fascia. Instead, there are numerous movement modalities that utilise various fascial properties to different degrees in a (more or less) deliberate manner.

While some fascia-focussed training concepts are more versatile, engaging fascia in many different ways, others are more specialised; for example, by focussing on elasticity with dynamic, oscillating exercises, or plasticity changes with melting stretches, or stimulation with self-massage.

Slings Myofascial Training aims for diversity. Because we strive to utilise and improve a wide range of fascial qualities, the movement repertoire and sequencing are versatile and contrasting.

When it comes to the Slings exercise application, the focus is very much on a cognitive and somatic understanding of why we do what we do. When understanding the 'why', the 'how' of movement execution and sequencing becomes clear.

All movement is myofascial

Not all modalities train fascia deliberately

Know 'why' to determine 'how'

6 GUIDING PRINCIPLES

The Slings concept has six guiding principles, which you can view as the 'solid' elements that provide the structural support for the education and the practice itself. What links them are the interconnected twelve Fascial Movement Qualities. Together, they create the unique and adaptable shape of Slings Myofascial Training, in which one element supports and creates change in all of the others.

The Slings Myofascial Training concept is:

1. Experience-based
2. Science-informed
3. Resource-oriented
4. Curiosity-driven
5. Diversity-embracing
6. Ignorance-conscious

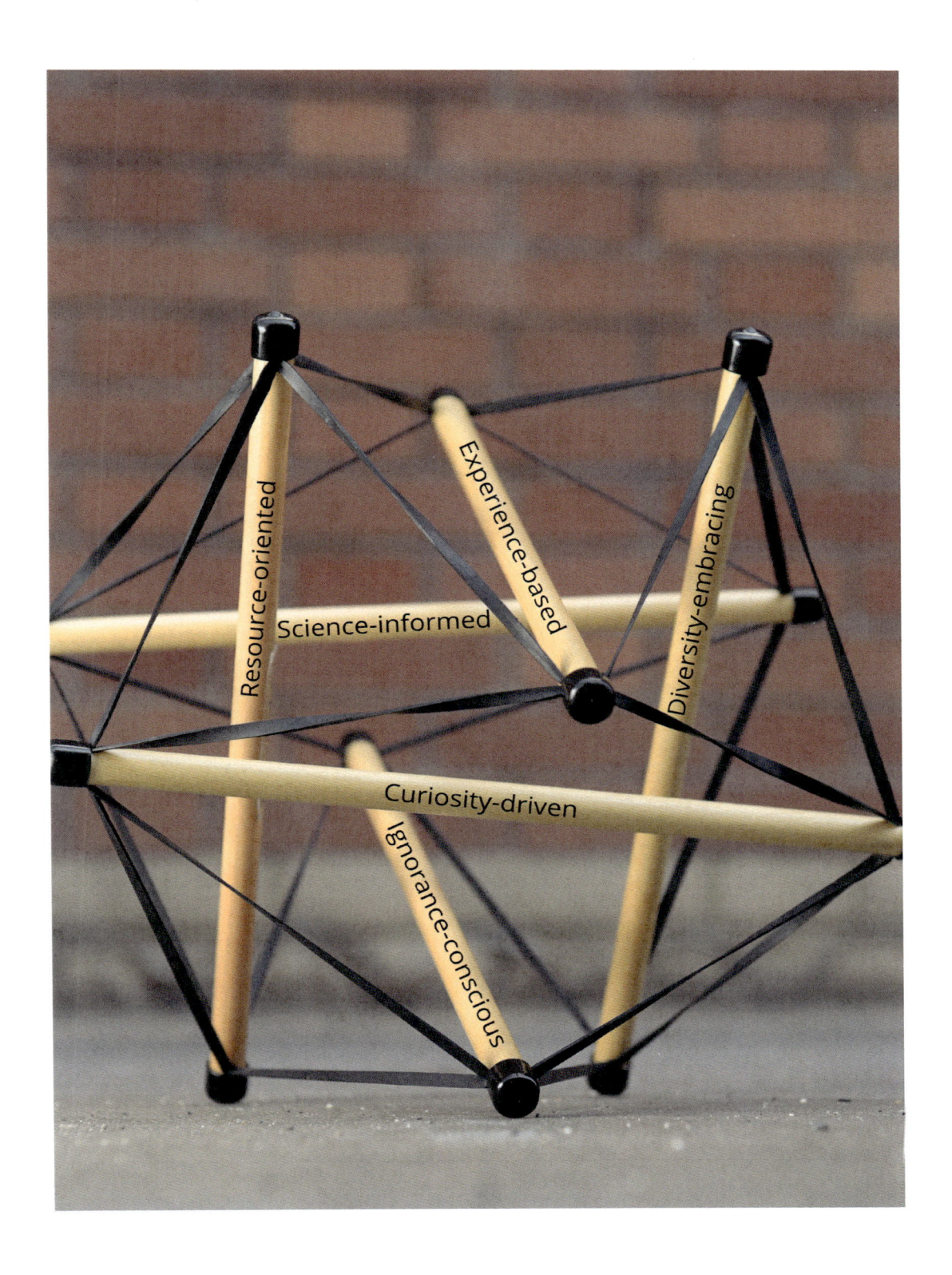

Slings Concept

1. Experience-Based
The refinement of exercises and sequencing through diligent practise

If it works in practice, it works

Personal practice, teaching, and observation have considerably shaped the development of the Slings Myofascial Training concept. They have shown what works regularly, occasionally, and rarely. Slings-certified teachers, art of motion educators, and my growing experience continuously spark my curiosity to know more, understand better, explain more comprehensively, and explore further. This journey of ongoing exploring, questioning, achieving, failing, and succeeding certainly made me – and I believe the teachers as well – a little humbler and a little wiser.

The way I see it: if it works in practice, it works. Experience is non-negotiable, by science or by people whose experience is different. If it doesn't work, tweak it until it does and learn from the process. As you gain practical experience, ponder the 'why' and let yourself be informed by science and other credible sources.

2. Science-Informed
The translation of cognitive insights into practical application

Think, question, translate

The well-structured approach and often visionary work of scientists, researchers, and individuals searching for answers have significantly informed Slings Myofascial Training. Besides increasing my understanding, curiosity, and questions, it helped me formulate better questions and find clearer answers.

The way I see it: research can give invaluable clues. It invites us to think, question, 'translate' scientific information into practical application, reflect on outcomes, stay curious, and ask more specific questions.

3. Resource-Oriented
Recognising and strengthening the body's resources to feel internally empowered: in health and illness

Strengthen trust

In Slings, we focus on physical, cognitive, and emotional resources and the resourcefulness with which they can be used. This doesn't mean deficiencies, weaknesses, or illness are ignored. Instead, the goal is to address them with a greater sense of somatic courage, strength, and resilience to facilitate self-healing.

The way I see it: a resource-oriented approach strengthens the trust in one's own self-healing capacity, which in itself is healing. It is movement medicine with health in mind.

4. Curiosity-Driven
The subject matter knowledge and courage to keep looking for what is not yet seen

Curiosity has been the initiator and remains one of the primary motivators for developing and remodelling Slings Myofascial Training. Curiosity leads to exploration, and exploration to new experiences. New experiences can lead to new discoveries, the occasional dead end, insightful failures, and of course, small and large successes!

The way I see it is, as Einstein said: "The important thing is not to stop questioning. Curiosity has its own reason for existing. One cannot help but be in awe when he contemplates the mysteries of eternity, of life, of the marvellous structure of reality. It is enough if one tries merely to comprehend a little of this mystery every day. Never lose a holy curiosity".

Never stop asking questions

5. Diversity-Embracing
The ability to appreciate and support uniqueness

Human beings, their bodies, emotional states, and learning styles are diverse. Each person is unique in her or his neuro-myo-fascial-skeletal-psycho-emotional-social-perceptual composition, something I believe should be acknowledged and reflected in movement practice. A 'one for all' approach negates the diversity that makes us human (and interesting!). Slings is a concept that needs to be understood in its essence so it can be customised to the different postural patterns, movement skills, needs and wants, and last but not least, the unique personalities of the movers. The ideas, principles, and exercises Slings contains are not meant to be merely memorised or imposed on bodies and psyches.

The way I see it: the uniqueness of our body, mind, and spirit are qualities worth cultivating instead of negating or worse eliminating.

Honour uniqueness

6. Ignorance-Conscious
The self-confidence to acknowledge one's own physical, intellectual, and emotional unawareness

In the first few years, developing Slings felt like assembling a puzzle without having all of the pieces, seeing the complete picture, or knowing if there even was a complete picture. Over time, I have collected and united elements into what I believe to be a coherent whole that will keep evolving. It has been exciting and hard work, full of learning, idea-doubt, rewards, and times of 'thoroughly conscious ignorance', which keeps me grounded and progressing.

The way I see it: acknowledge your skills and knowledge because it is healthy and feels good. Acknowledge your ignorance because it keeps you humble and supports learning.

Stay humble

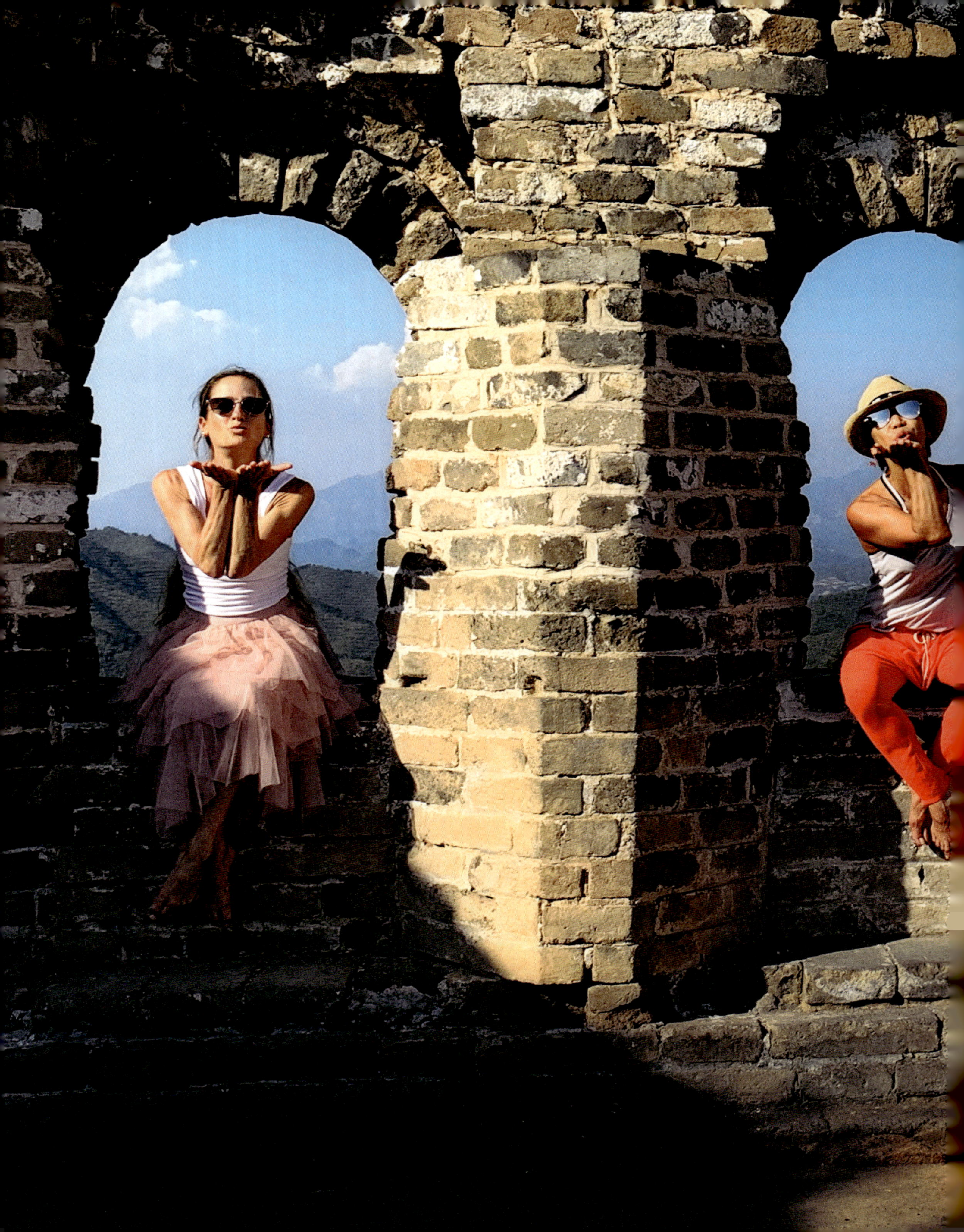

PART 2
SLINGS
MYOFASCIAL TRAINING TRINITY

CHAPTERS OF PART 2

12 Training Techniques

12 Training Aims

12 Fascial Movement Qualities

SLINGS MYOFASCIAL TRAINING TRINITY

The twelve Slings Fascial Movement Qualities defined in this book serve as the basis for the twelve Slings Myofascial Training Techniques. The techniques are practically applied to achieve specific exercise goals that are coherent with the defined fascial qualities. In sum total, the benefits of the individual exercises facilitate the embodiment of the twelve Slings Training Aims.

12 Fascial Qualities

1. Tensile Strength
2. Muscle Collaboration
3. Force Transmission
4. Adaptability
5. Multidimensionality
6. Fluidity
7. Glide
8. Elasticity
9. Plasticity
10. Tone Regulation
11. Kinaesthesia
12. Imponderability

12 Training Techniques

1. Stabilising
2. Toning
3. Pushing, Pulling, Counter-traction
4. Active Lengthening, Expansion
5. Spiralling
6. Hydrating
7. Gliding
8. Domino Motion
9. Bouncing and Swinging
10. Massaging
11. Active Ease
12. Melting and Invigorating

12 Training Aims

1. Dynamic Stability
2. Multidimensional Strength
3. Limberness
4. Movement Rhythm
5. Elasticity
6. Tissue Nourishment
7. Adaptability
8. Resilience
9. Somatic Trust
10. Movement Courage
11. Mañana Competence
12. Kinaesthetic Intelligence

PART 2

12 FASCIAL MOVEMENT QUALITIES

Inherent, distinguishing features relevant to movement

A quality describes the characteristics of fascia. In our context, these characteristics relate to movement. In that sense, the twelve Fascial Movement Qualities describe inherent and distinguishing features of fascia that relate to their roles and accomplishments for the body in motion. Here they are, the twelve Fascial Movement Qualities we deliberately work with in Slings.

1. Tensile Strength
2. Muscle Collaboration
3. Force Transmission
4. Adaptability
5. Multidimensionality
6. Fluidity
7. Glide
8. Elasticity
9. Plasticity
10. Tone Regulation
11. Kinaesthesia
12. Imponderability

1. Tensile Strength	**2. Muscle Collaboration**	**3. Force Transmission**
4. Adaptability	**5. Multidimensionality**	**6. Fluidity**
7. Glide	**8. Elasticity**	**9. Plasticity**
10. Tone Regulation	**11. Kinaesthesia**	**12. Imponderability**

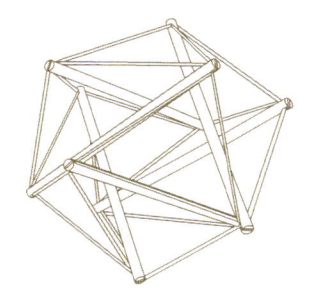

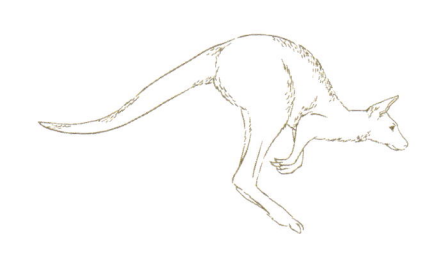

THE STORY OF THE ICONS

Each icon represents attributes of the respective quality. A brief explanation is provided for understanding the symbolism.

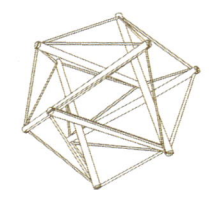

Tensile Strength
The elastic bands of the tensegrity model represent the tensile strength of fascia. The model as a whole illustrates the body, which – to a considerable degree – is dynamically stabilised by the inherent tension of the fascial web.

Muscle Collaboration
The well-developed biceps brachii depicts a well-conditioned muscle. The arm itself represents the importance of collaborative muscles for optimal fascial functionality.

Force Transmission
The connected hands illustrate myofascial links. The handshake represents the force transmission between adjacent fascial structures.

Adaptability
The chameleon, whose skin colour changes for specific reasons, illustrates the lifelong ability of the fascial system to adapt to short and long-term purposes.

Multidimensionality
The various shapes, structures, textures, and densities of the components of an orange represent the multidimensionality of the fascial system.

Fluidity
The water droplet illustrates fascial water, the fluid nature of fascia, and fluid flow in fascia.

Glide
The smooth swimming fish represents well-lubricated glide between fascial layers.

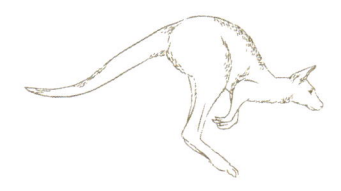

Elasticity
The kangaroo, with its elastic jumping power, depicts the ability of fascia to elastically store and release energy.

Plasticity
The burning candle in which the wax melts, changing the form of the candle, illustrates fascial plasticity, the tissue's ability to reshape under specific conditions.

Tone Regulation
The strong rope depicts fascial tone. The knot illustrates contractility and an increase in tissue firmness. That a knot can be done and undone represents the fascial system's ability to regulate its tone.

Kinaesthesia
The figure 6 stands for fascia being the sixth sense. Inspired by Oliver Sack's phrase "kinetic melody", the notes represent kinaesthesia. That the notes are within the figure 6 depicts that kinaesthesia occurs within the fascial system.

Imponderability
The cat stands for curiosity and the sense of wonder that comes with the imponderability of fascia as a holistic system.

PART 2

Distinguishing Features

Before looking at each movement quality in detail, here is a different kind of overview of the distinguishing features and roles of fascia in motion.

1. Fascia is **tensile**; therefore provides bodywide dynamic stabilisation.

2. Fascia is **collaborating with muscles**; therefore, all movement is myofascial.

3. Fascia is **force-transmitting**; therefore is a mechanical communication system.

4. Fascia is **adaptable**; therefore, its architecture shapes and remodels through movement.

5. Fascia is **multidimensional**; therefore permits and requires multidimensional movement.

6. Fascia is **fluid**; therefore provides a liquid, health-promoting habitat for cells.

7. Fascia is **gliding**; therefore enables sliding motions between adjacent myofascial structures.

8. Fascia is **elastic**; therefore has the capacity to contribute energy to movement.

9. Fascia is **plastic**; therefore contributes to the shape of the body and movement.

10. Fascia is **tone regulating**; therefore, its firmness is variable.

11. Fascia is **kinaesthetic**; therefore, a rich sensory system.

12. Fascia is **imponderable**; therefore, greater than the sum of its parts.

PART 2

The **tensile strength** of fascia
supports your postural and movement ergonomics,
makes you resilient, and unloads your body.

1. TENSILE STRENGTH

Fascia is a continuous tensile communication network and an integral part of the body's tensegrity. Tensegrity is the structural explanation for bodywide dynamic stability, responsiveness, and spaciousness.

Let Me Ask You
How do some people retain the health of their joints into old age? What differentiates an upright posture from the postural poise that catches your eye? Tensegral balance.

Fascia is inherently tensile. As a bodywide, interconnected network it provides tensile strength from head to toe. Together with bones, it forms a tensegrity system that supports body balance and spaciousness, along with natural grace.

A Good Reason To Care
If eye-catching elegance is not a good enough reason to get onto the mat and care about balanced fascial tension, the longevity of your joints and organs certainly is!

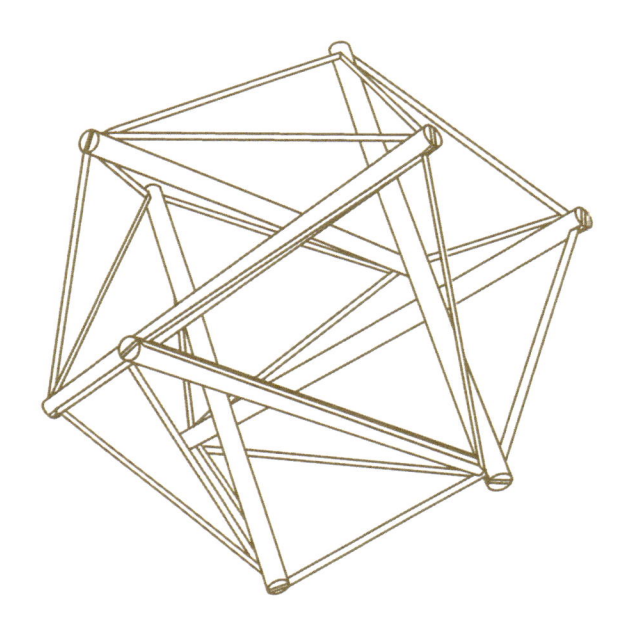

PART 2

BALANCING TENSION AND EXPANSION

"Tensegrity structures can be seen as restrained expansion. Expansion (or space) creates tension."
                                                        Stephen M. Levin

TENSEGRITY

Tensional integrity

Tensional and integrity come together to form the word tensegrity, which finds its roots in art and architecture. In the middle of the 20th Century, the American artist and photographer Kenneth Snelson created world-famous pieces of art. The pieces beautifully depicted what was later termed tensegrity by Richard Buckminster Fuller, a renowned architect, systems theorist, author, designer, and inventor. Snelson himself used the description 'floating compression'.

Donald Ingber of Harvard University was the first to apply this architectural concept to the body. While Ingber pointed to tensegrity at the cellular level, Steven Levin went on to indicate its presence in gross structures of the body, not just in cells. Especially the pelvis, shoulder girdle, and knee joints caught his attention. However, Levin proposes that all of the body expresses biotensegrity, or what we call 'tensegral qualities'.

There is still debate about exactly how the body lines up with human-made tensegrity structures, recognising it does not have the same kind of linear responses. Nevertheless, biotensegrity provides a useful model for identifying balance or lack of balance in the body, as well as describing a certain movement quality that will be discussed later in this chapter. By applying the theory of tensegrity, the centuries-old concept of the skeleton as a frame upon which soft tissue is draped is evolving. The body is recognised as an integrated system in which bones are suspended in a continuous network of fascial tension.

Discontinuous elements are suspended in a tensional network

Stable, adaptable shape

In simple terms, a tensegrity system consists of discontinuous compression elements that are suspended in a continuous, non-linear tensional network. Just like the body, a tensegrity system stabilises itself without external support and maintains its shape independent of gravity.

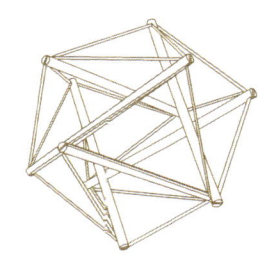

Tensegral

It is worthwhile for fascia-focussed practitioners and movers to add the word tensegral to their movement vocabulary. Derived from the word 'tensegrity', it describes the qualities of a system or action that are coherent with the functional aspects of a tensegrity system.

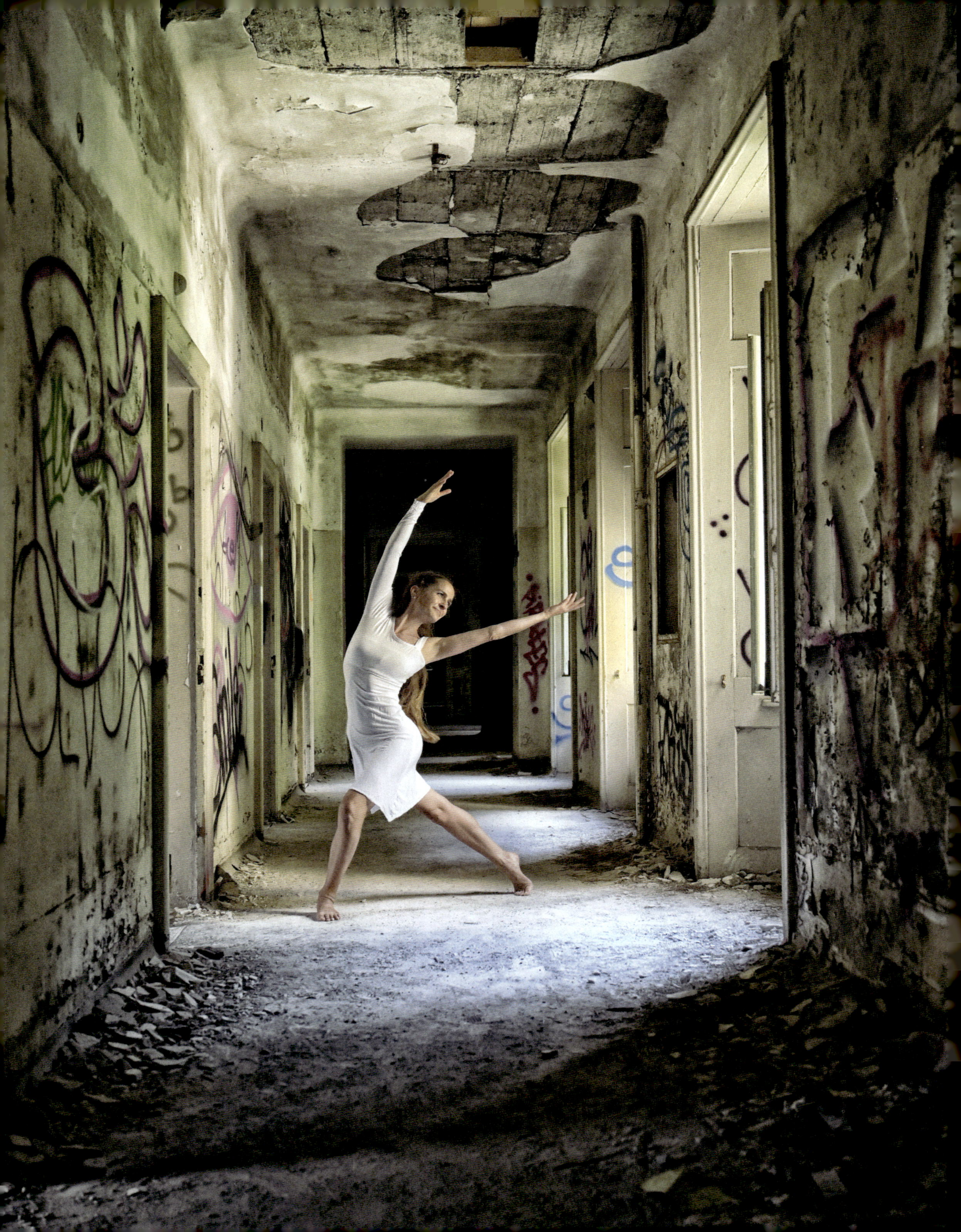

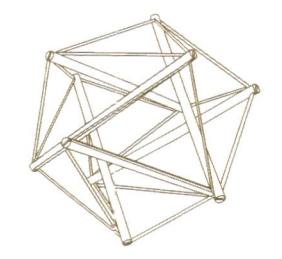

Tension and Compression

Structures are built to withstand loads. This is true for more static constructions (buildings, towers, and bridges) and those built to move (airplanes, vehicles, and bodies). In simple terms, to measure a structure's capacity for load, one can look at compression strength and tensional strength. Compression strength is a material's ability to resist forces pushing in (think wooden rods, bricks, steel girders, or cartilage). Tensional strength is a material's capability to resist forces that pull apart (think elastic bands, cables, guy wires, or fascia). In man-made compression structures, like walls, pillars, and bridges, compression members are assembled together part by part, brick by brick, and stone by stone. In essence, they press against each other. In tensegrity structures the floating compression elements resist out against the tensional network that pulls everything into a resilient harmony. This interaction of compression and tension develops health in our system as we grow; and continues to sustain health for a lifetime, particularly when challenged through active, healthy living.

By describing the body as a compression structure with levers, biomechanics fell significantly short in describing our magnificent inner design that can sustain longevity into old age. Tensegrity is much more appreciative and inclusive of our inner complexity. However, it is important to remember that we are still using a man-made construct to explain the holistic, imponderable nature of the body in motion.

SOMATIC TENSEGRITY

Fascia is inherently tensile, similar to a rubber band that is never slack but stretched to various degrees of tautness. Together with the bones, fascia forms a tensegral system within your body. In your somatic tensegrity, the discontinuous elements (bones) are suspended in a continuous tensile network (fascia), without touching each other (joint spaces). Looking at it the other way around, the discontinuous elements (bones) push out, creating tension in the continuous tensile network (fascia), which sustains the spaciousness of the whole system (body). Embedded in the body's tensegrity are also actively contractile elements, or 'tension-tuners' (muscles). Their tone can increase or decrease fascial tension, therefore increase or decrease stability or adaptability, respectively.

Bones, fascia, and joint spaces are dynamically balanced

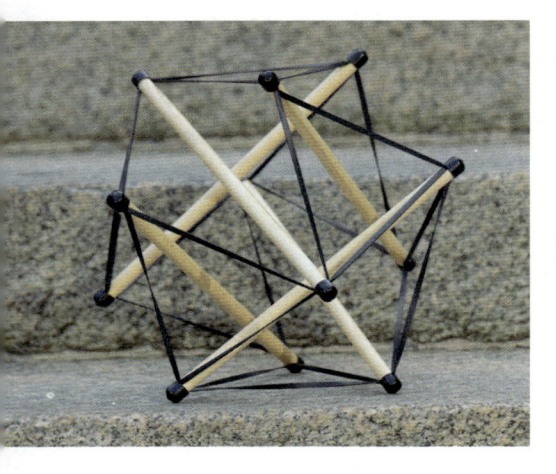

Bones and Fascia in Tune

The tensegrity model shown in the photograph represents the body.

Wooden sticks: The wooden sticks represent the bones, the discontinuous compression elements.

Elastic bands: The elastic bands represent fascia, the continuous, non-linear tensile network.
Within the elastic bands, you can imagine red muscles, the actively contractile tension-tuners of the body's tensegral system.

Spaces: The spaces represent joints, physical spaciousness.

Sturdiness amidst pliability facilitates resilience

Sturdy, yet pliable, tensegrity structures are resilient, a quality that is desirable in the body. Unlike a tensegrity model, though, bodies are wonderfully unique. The discontinuous and continuous components vary in length, shape, composition, orientation, and where they are relative to gravity and each other. Also, in the moving body, fascial tension alters constantly.

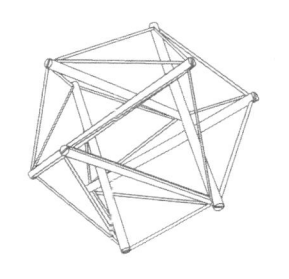

Muscular Fine-Tuning

Muscle activity influences the tensional equilibrium of fascia. How much muscle tone is 'right' for your body depends on the tensile strength of your fascia, your activities, and the way you load and unload your somatic tensegrity. Naturally, body composition and functional demands change throughout life. Staying 'in tune' is a lifelong project.

The Slings concept offers a broad training spectrum that engages muscles lightly, maximally, and everywhere in between for various periods of time and at different speeds. The muscular system is more responsive and attuned to internal and external changes when it is well-conditioned, both inside and out.

Muscles fine-tune fascial tension

QUALITIES OF SOMATIC TENSEGRITY

Somatic tensegrity has qualities that are worthwhile to consider in a fascia-focussed movement practice.

Dynamic Stability

The evenly distributed tensile strength of a tensegrity system provides energy efficient stability, therefore assists structural integrity.

Resilient Adaptability

A balanced tensegrity structure is resilient. Meaning, it can be temporarily deformed in a multidirectional manner through tensional or compressive forces, without being damaged. Once unloaded, the system effortlessly recovers its original shape.

3-Dimensional Expansion

A tensegrity structure is three-dimensionally expansive. When two elements move apart, the whole system expands.

Bodywide Communication

During movement, the stress placed on one part of the system is dispersed widely. In this way, tensional changes anywhere are instantly signalled to everywhere else in the body. This interrelated arrangement facilitates a whole-body response to local events.

Symptoms and Sources

The bodywide mechanical communication provided by tensegrity is an excellent function that unloads individual structures. However, it can also present a somatic challenge when trying to locate the source of symptoms. Feelings such as pain can emerge in one part of the body – perhaps a weak place in the system – yet their origin might be elsewhere.

FUNCTIONAL FASCIAL TENSION

Dynamically stabilising and functionally adapting

Influenced by genetic predisposition, lifestyle, activities, somatic demands (and more), fascial tension varies in the body and between bodies. Some people's inner tensegrity is more stable, while others have more pliancy. There is no right or wrong degree of tension. What there is, though, is a degree of tensile strength that supports both dynamic stability and adaptability, which promotes healthy joint alignment and range of movement. In that sense, the training aim is to sustain functional fascial tension that adapts to physical demands; immediately and with the natural fluctuations of living life.

Dysfunctional Fascial Tension

Fascial rigidity: Leads to joint fixation and internal compression, decreasing structural adaptability, and spaciousness.

Fascial laxity: Leads to joint instability or muscular compensations as well as internal compression, decreasing structural integrity, and spaciousness.

BREATH AT ITS CORE

3-dimensionally expansive

At the core of somatic tensegrity is the breath. An inhalation three-dimensionally expands the thorax and elongates the spine, which increases tension in the taut fascial elements. An exhalation is a gentle retraction that lessens tension in the taut fascial elements while tensioning other fascia with altered intraabdominal pressure. In an adaptable tensegrity system, respiration supports local and bodywide dynamic stability as well as physical expansiveness.

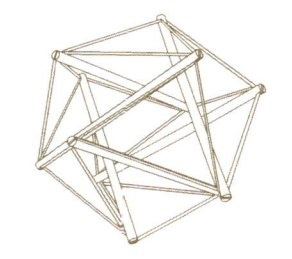

IS ALL OF THIS TRUE?

You might wonder if all of this is true. What about the concept of levers and fulcrums we studied so extensively in biomechanics classes?

Tensegrity Versus Lever System
"The body is not (unless injured or misused) the 'strain focussing machine' described in biomechanical texts, but is rather a 'strain distribution system' with myofascial meridians providing pathways for the body's tensegrity."
Thomas W. Myers

It is evident that viewing the body as a biological tensegrity system is quite different from the lever mechanics postulated by Giovanni Alfonso Borelli in 1680. Stress placed on a lever system creates a local impact. In terms of the body, this means individual joints are compressed. In contrast, stress placed on a tensegrity system is distributed through its entirety. In terms of the body, this means that strain is distributed throughout the myofascial-skeletal system, taking stress off individual joints.

Practice-Proven Theory
There is support within the scientific community that tensegrity holds true on a cellular level, as shown by Donald Ingber. However, there is currently no agreement concerning somatic tensegrity. What seems somatically clear is that the body has (or can have) tensegral qualities. Viewing and sensing the body in this way changes the perspective from seeing yourself and others as an aggregate of more or less functional parts, to recognising the body as a single organism with interconnected components.

Scientifically validated or not, the body has tensegral qualities

Viewing the body through the lens of tensegrity has changed my movement priorities; it opened my eyes to postural and movement possibilities that I didn't even see before. Until I learn otherwise through science or experience, I will continue to work with the concept of somatic tensegrity, because it is conceptually illuminating and practically useful. If it resonates with you, I encourage you to work with it and explore its possibilities.

Which One Would You Rather Be?
Strain-focussing lever system or strain-distributing tensegrity system?

SLINGS IN MOTION

DYNAMIC STABILITY AND CENTEREDNESS

Dynamic stability and centeredness exist on a spectrum

Postural ease

A balanced somatic tensegrity facilitates optimal dynamic stability and centeredness. This means well-aligned bones and evenly distributed myofascial forces. Keep in mind that both dynamic stability and centeredness exist on a spectrum; they are not fixed points at which the body is held in 'the' neutral position. When the body is dynamically stabilised and centred, posture is elongated in an easeful manner.

To make the concept of dynamically stabilised centeredness more tangible, visualise it as a scale. The extreme ends of the scale are instability and fixation. In the middle is the space in which equilibrium is dynamically maintained with minimal effort. In Slings, we call it an active resting place.

Instability

Structural disintegration

Instability refers to the inability to maintain a beneficial joint position posturally or in movement. This excessively strains joints and associated soft tissues.

Myofascial Involvement

- Muscles can be hypotonic due to weakness or inhibition.
- Fascia can be generally too lax and/or insufficiently tensioned when moving.

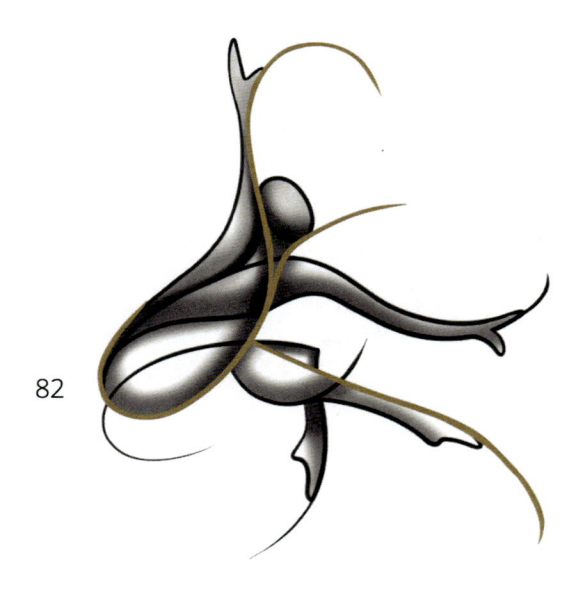

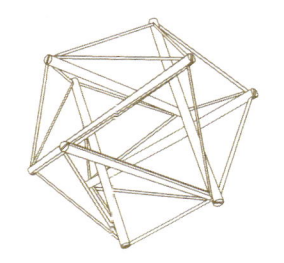

Fixation

Fixation refers to a muscle and fascia induced joint immobility that restricts a healthy range of movement, increases pressure within joints, and eventually leads to compensation patterns.

Structural rigidity

Myofascial Involvement

- Muscles can be hypertonic due to chronic, habitual, or compensatory over-engagement.
- Fascia can be too stiff, felty, or adhered.

Instability   Centeredness / Dynamic stability   Fixation

Equilibrium Through Movement Variety

Dynamically balanced movement diversity

One of the primary aims of Slings Myofascial Training is to rebalance and maintain a dynamic equilibrium in the body's tensegrity. It is not just one type of movement or one kind of training technique that does the magic, though; it is the cumulative effect of diverse exercises. The creative challenge is to find an appropriate balance of toning, lengthening, hydrating, gliding, stimulating, and calming motions. Naturally, the perfect one-for-all recipe doesn't exist. This may sound a bit surprising, but speaking from experience, incorporating a little bit of everything works exceptionally well. With your own experience, you will fine-tune the blend and ratio.

TENSEGRAL MOTION: ACTIVE EASE

Healthy opposition performed with ease

Tensegral motion has an intrinsically polarised quality. Regardless of the direction, there is always a sense of counter-traction, drawing in and reaching out, spiralling and expansion. The polarity of tensegral motion provides continuous tensional support, which sustains the spaciousness of the body. It is performed with a sense of active ease during which fascia is tensioned and softened without excessive force or collapse.

Karin and Lucas arching tensegrity-style at the main station in Berne in Switzerland

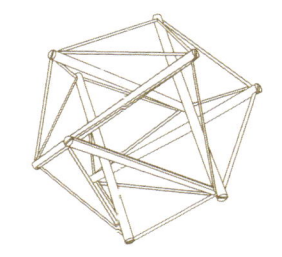

Active Lengthening
Active lengthening of myofascial structures stimulates fibroblasts and tensions muscle fascia as well as adjacent deep and loose fascia. This makes it a highly effective form of fascial conditioning. It also supports the tensegral balance between contralateral fascial structures, especially in and around the spine. Therefore, conscious reaching and counter-traction (when possible) are incorporated in active lengthening exercises.

Strength in length

Ribcage Expansion
In Slings, there is a great deal of focus on three-dimensional ribcage expansion. This is done directly with the breath by emphasising expansion with the inhalation and gentle retraction with the exhalation. It is also indirectly supported by increasing glide and adaptability of the fascia in and around the ribs.

Active expansion, relaxed retraction

Roll Down

Side Bend & Rolling with Massage Balls

PART 2

3-dimensional helical motion

Spiralling

A rotation occurs around an axis and remains on one plane. A spiral is helical; it coils around an axis in a constantly changing series of planes. You can view a spiralling motion as a three-dimensional rotation. Because of their tensegral nature, spirals are frequently incorporated in exercises for the spine, ribcage, shoulders, and hips.

Spine

Spinal spirals are a favourite element of the Slings repertoire! Because of its architecture and relationship with the breath, the thoracic spine is in focus, particularly the mid-thoracic area (T2-T9), where the ribs can act as spacers between the two vertebrae they contact. In spiralling spinal exercises, the breath is consciously incorporated and used to increase movement and structural dimensionality in and around the spine.

In spinal spiralling exercises, the breath is consciously incorporated and used to increase dimensionality.

- Inhalation: Because the inhalation elongates the spine, spinal spirals are commonly executed with a breath in.

- Exhalation: Because the exhalation naturally creates gentle retraction, this phase offers itself for elastic recoiling or a sense of softening.

Spiralling Twist & Curl with Figure-8 Arm Sweep

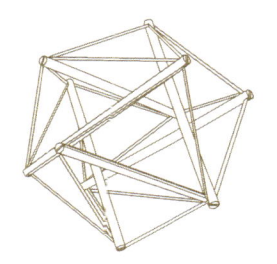

Ribcage

The ribcage is a beautiful example of tensegrity in and of itself. The structure's expansiveness and its spiralling dance with the spine encourage movement freedom, ease in the shoulders and neck, and allow you to take full breaths effortlessly. This dance incorporates gentle spiralling of the entire ribcage, as well as the individual ribs rotating within their fascial envelopes.

Although spiralling of the elongated spine facilitates rib mobility beautifully, practical experience has shown the value of a different kind of multidimensional motion. When the spine is taken into a side bend, then rotated, the ribcage spiralling is enhanced. Combined with deep breaths, this effectively and joyfully supports inner tensegrity in the moment and thereafter.

Cleopatra with Spiralling

Extremities

By reaching out or drawing in during rotational movements of the legs or arms, we also encourage spiralling in the hip joints and shoulder joints, even if it is just by intention. From the shoulders, spiralling can continue out into the arms (radioulnar joint) and hands.

Kneeling Side Support with Leg Lift &
Leg Kick Front with Spiralling

… ONTO THE MAT AND INTO THE BODY

▶ Story and Practice: Tensile Strength in Motion

SLINGS MYOFASCIAL TRAINING TECHNIQUES AND AIMS

In the following lists, which repeat for each of the twelve Fascial Movement Qualities chapters, the fonts differentiate immediate relevance.

The Slings Myofascial Training Techniques shown in the gold, bold font are in focus and considered to have a direct impact on the discussed Fascial Movement Quality of the chapter. Regular font indicates that these techniques play a role but are not our direct focus. The lighter shaded items have little or no direct conceptual relevance at this point.

In the table on the right, identified in gold, bold font, are the top four Training Aims most relevant to the focus of each chapter.

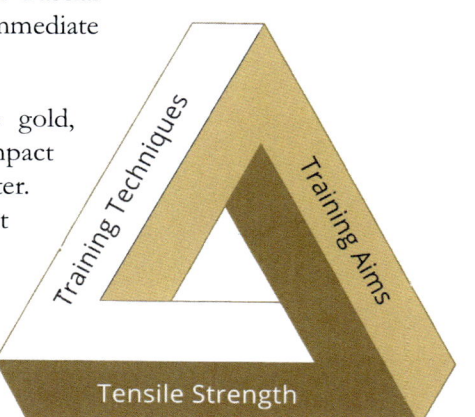

12 Myofascial Training Techniques
Techniques that directly influence tensegral balance:

1. **Stabilising**
2. Toning
3. Pushing, Pulling, Counter-traction
4. **Active Lengthening and Expansion**
5. **Spiralling**
6. Hydrating
7. Gliding
8. Domino Motion
9. Bouncing and Swinging
10. Massaging
11. Active Ease
12. Melting and Invigorating

12 Training Aims
Four top achievements of tensegral balance:

1. **Dynamic Stability**
2. Multidimensional Strength
3. Limberness
4. Movement Rhythm
5. Elasticity
6. Tissue Nourishment
7. Adaptability
8. **Resilience**
9. **Somatic Trust**
10. **Movement Courage**
11. Mañana Competence
12. Kinaesthetic Intelligence

PART 2

IN A NUTSHELL
THINGS WORTH KNOWING ABOUT FASCIAL TENSILE STRENGTH AND MOVEMENT

4 Things About (Bio)Tensegrity

Tensile fascial network:	Fascia is inherently tensile. The degree of tension naturally varies in different body parts and people.
Somatic tensegrity:	Somatic tensegrity consists of discontinuous elements (bones) that are suspended in a continuous, non-linear tensile network (fascia), without touching each other (joint spaces). Embedded are actively contractile 'tension-tuners' (muscles).
Resilient form:	Tensegrities are resilient, energy-efficient systems that maintain or change and regain their form regardless of orientation and gravity.
Tensegral equilibrium:	Tensegral equilibrium in the body is characterised by balanced tensile (fascia) and contractile (muscles) forces that organise and dynamically stabilise the discontinuous elements (bones).

4 Things About Expansion and Dynamic Stability

3-dimensional expansion:	A tensegrity structure is three-dimensionally expansive. When two elements move apart, the whole system expands.
Bodywide dynamic stability:	The continuous tensile strength of fascia provides bodywide dynamic stability on a structural level.
3-dimensional breath:	Breath is a key contributor to three-dimensional expansion and dynamic stability in somatic tensegrity.
Muscular fine-tuning:	Muscle tone and activity influence the tensional equilibrium of fascia. As a result, purposeful muscle conditioning can fine-tune the tensile balance of fascia.

4 Slings Training Principles

Intrinsically contrasting:	Tensegral motion has an intrinsically polarised quality. Regardless of the direction, there is always a sense of elongation and spaciousness in the body.
Active ease:	Tensegral motion is performed with a sense of active ease in which fascia is tensioned and softened without excessive force, compression, or collapse.
Length, expansion, spirals:	Tensegral motions often include active lengthening, ribcage expansion, and spiralling movements.
Breath:	The consciously incorporated breath adds both dimension and dynamic stability to all movements.

3 BENEFITS OF FASCIAL TENSILE STRENGTH

1. Bodywide dynamic stability

2. Physical resilience

3. Structural longevity

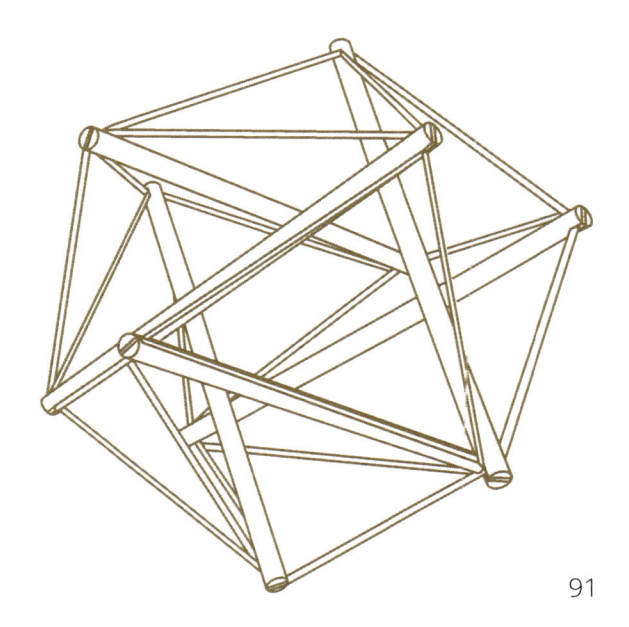

PART 2

The **muscles** move and empower your fascia.

2. MUSCLE COLLABORATION

Muscles are movement motivators, embedded in and functionally linked with fascia; every movement is myofascial. For dynamic balance between the muscular and the fascial system, muscles need to be strong, flexible, and able to relax.

Let Me Ask You
Does regularly lengthening and strengthening your muscles feel good? It sure does!

Muscles are amazing. They are stabilisers, movers, energisers, heaters, and protectors. Very importantly, they are the other half of the 'myofascial' synergy, inseparably linked to fascia in motion. Your body's functionality relies on their harmonious interplay.

A Good Reason To Care
The dynamic balance of your fascial system is one human-sized reason to get onto the mat and care for your 650-plus muscles regularly.

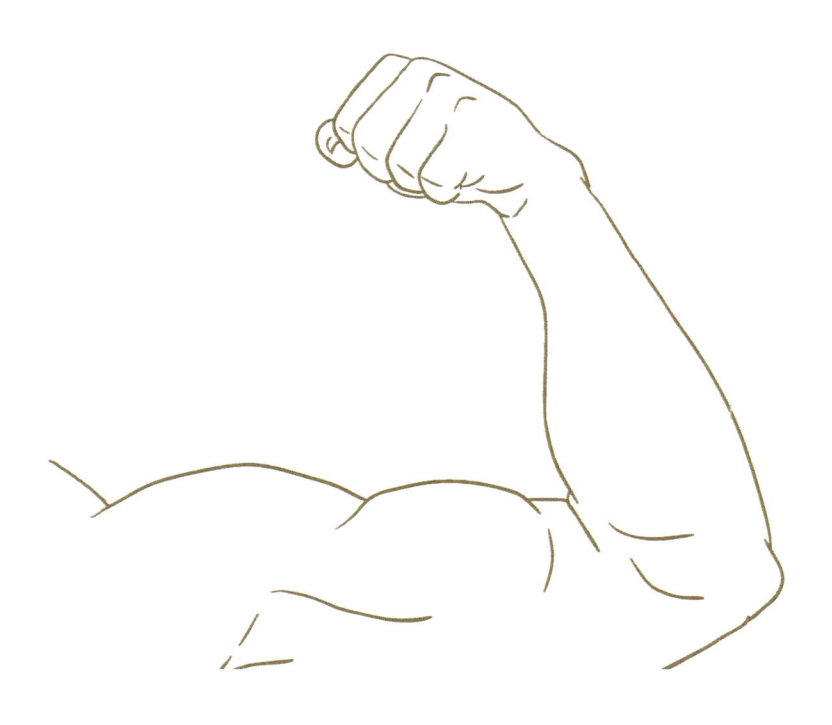

PART 2

(EM)POWERED BY MUSCLES

"Haruki Murakami, a Japanese bestselling novelist writes: 'The fact that it takes character to get out of the chair, is perhaps the greatest benefit to be derived from exercise.' Exercise does not simply make Murakami a better sportsman or writer, by enhancing his mind or enhancing his stamina. It also strengthens his 'muscles' of consistency and integrity, so his grasp of his own character is firmer for longer."
Damon Young

Source of motivation, stability, energy, groundedness

Fascia is many things for muscles: a connector and separator, a force transmitter and amplifier, a supporter and informer, as well as a sliding surface. For good reason, fascia has become quite famous in recent years. Should muscles now stand in the shadow of the glorious, occasionally glorified fascia? Absolutely not. Fascia doesn't go anywhere without muscles. They are a major source of motivation, both literally and figuratively, enabling you to use your body wilfully, purposefully, and creatively. Muscles are the home of the body's own energisers, the mitochondria, and act as 'fine-tuners' for fascial tension. Muscular contributions to local joint stability and a healthy sense of being grounded are central to all-around wellbeing. Because movement is a synergy of muscles and fascia, improving functionality is not about excluding one or the other, but rather integrating fascia-focussed and muscle-focussed exercises consciously into the practice.

Balanced interplay between muscles and fascia

Nik and Karin feeling and being empowered by muscles in an industrial area in Berne in Switzerland

95

PART 2

SLINGS IN MOTION

VERSATILE MUSCLE CONDITIONING

The Slings repertoire includes exercises for strengthening muscles in all kinds of ways, lengthening them in active and relaxed manners, and of course, supporting the ability to relax.

Differentiated Strength

Differentiation of core stabilisation from general muscle toning

To match the specificity of Slings in an ideal way, core stabilisation and general muscle toning exercises are differentiated from each other.

Detail about the technicalities of core stabilisation and globally-oriented training will be saved for another time because they are complex topics in their own right. Instead, the next few paragraphs serve as a general reminder of how important inside-out muscle conditioning is.

Dynamic Stability

Dynamic stability provided by local, posture-oriented muscles is vital for functional movement and structural integrity. For these and other good reasons, central and bodywide core stabilisation exercises are deliberately woven into a Slings lesson.

3-dimensionally adaptable joint support

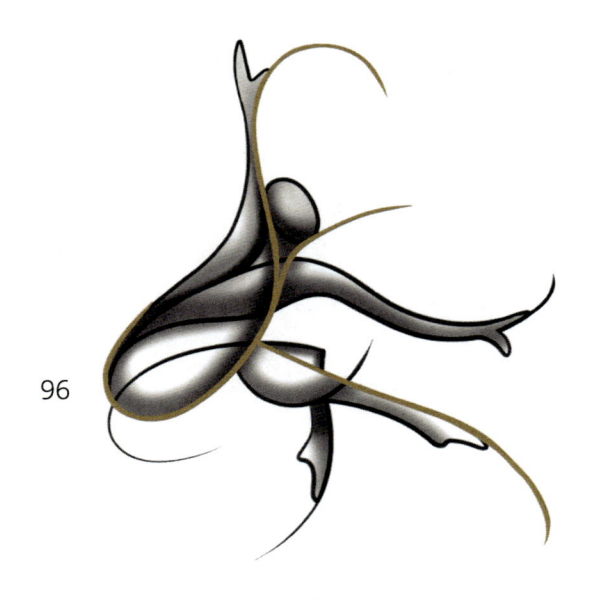

The word 'dynamic' means variation and contrast in force or intensity. In this sense, we see how it also means 'adaptability'. Well-functioning core stability does not fixate the body. Instead, it provides adaptable joint support that permits three-dimensional movement freedom, which needs to be reflected in core stabilisation exercises.

Central Myofascial Core

In Slings, the body's central core is called 'the Centre'. The term refers to the lumbar-pelvic region, where the body's centre of gravity is situated. The centre of gravity is a hypothetical place where the three planes of movement – the sagittal plane, frontal plane, and transverse plane – intersect. In an upright stance, it is located somewhere between the lowest lumbar vertebrae and the sacrum, in the middle of the pelvis. In the narrower definition, the Centre encompasses the pelvis and lumbar spine. In the more extensive definition, the whole spine is included.

The Centre

The muscles and fascial structures in focus are the:

- Pelvic floor
- Anterior sacral fascia and anterior longitudinal ligament
- Transversus abdominis
- Thoracolumbar fascia
- Multifidus
- Diaphragm

PART 2

The Deep Front Line

Bodywide Myofascial Core
The Deep Front Line of the Anatomy Trains concept is considered the bodywide myofascial core.

Relevé with Arm Arc & Plié with Arm Circle

Single Leg Balance: Knee Lift

Single Leg Plié

Single Leg Pivot

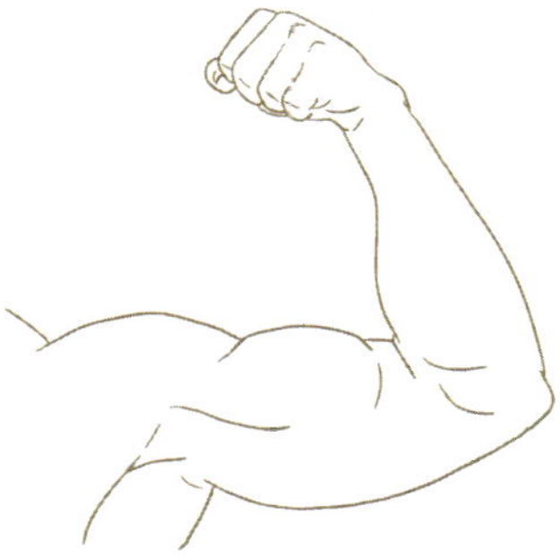

Multidimensional Strength

Everyday life benefits as much from multidimensional muscle strength as body-minded movement practices or athletic activities do. In Slings, we incorporate differentiated and integrated three-dimensional muscle conditioning exercises for the whole body. The repertoire also includes toning of global muscles around the Centre and the Deep Front Line. The overarching aim is to be and feel strong, regardless of the body position and life situation. What kind of exercises are performed and in which ratio depends on the answers to two questions.

For whom: Who are the exercises for?

For what: What are the exercises for?

Roll Down V Stance Balancing

90/90 Lunge

90/90 Lunge with Twist & Arm Reach

90/90 Lunge with Open Twist & Arm Reach

Inverted V Push Up: Out & Back

PART 2

BEING AND FEELING SUPPORTED BY MUSCLES

Preventing instability during fascial unwinding

The contribution of muscular strength to local joint stability and overall movement functionality is crucial during 'transformation phases' in fascia-focussed training. Transformation phases are periods of significant structural reorganisation when existing myofascial patterns are unwound, and new patterns are developed and integrated. During these times the body is more vulnerable to temporary, local instability.

Feeling supported and stable from within

Muscle toning can also bring about a sense of grounding and internal support. In other words, well-conditioned muscles can facilitate a sense of physical and emotional stability, which can dampen the internal 'noise' of change in progress.

Experience has shown the value and importance of incorporating muscle-focussed core stabilisation and general toning exercises throughout a Slings lesson. The more stimulating and unwinding the fascia-focussed exercises are, the more important it is to regularly rebalance and stabilise key areas like the pelvis and spine.

4-Point Kneeling Leg Press

Cat Series

SLINGS MYOFASCIAL TRAINING TECHNIQUES AND AIMS

The following Slings Myofascial Training Techniques and Training Aims shown in gold, bold font are directly related to intentionally engaging and conditioning muscles.

12 Myofascial Training Techniques
Techniques that focus on muscle conditioning:

1. **Stabilising**
2. **Toning**
3. **Pushing, Pulling, Counter-traction**
4. **Active Lengthening** and Expansion
5. Spiralling
6. Hydrating
7. Gliding
8. Domino Motion
9. Bouncing and Swinging
10. Massaging
11. Active Ease
12. Melting and Invigorating

12 Training Aims
Four top achievements of functional muscles:

1. **Dynamic Stability**
2. **Multidimensional Strength**
3. **Limberness**
4. Movement Rhythm
5. Elasticity
6. Tissue Nourishment
7. Adaptability
8. Resilience
9. **Somatic Trust**
10. **Movement Courage**
11. Mañana Competence
12. Kinaesthetic Intelligence

ONTO THE MAT AND INTO THE BODY

▶ Story and Practice: Muscle Collaboration in Motion

PART 2

IN A NUTSHELL
THINGS WORTH KNOWING ABOUT
MUSCLE COLLABORATION
AND MOVEMENT

3 Things About Myofascial Motion

Myofascial movement:	Muscles and fascia are an inseparable movement synergy; one cannot be engaged without the other.
Training differences:	Muscles and fascia have different training requirements.
Fascial functionality:	A well-conditioned muscular system supports fascial functionality.

3 Things About Dynamic Stability and Global Strength

Dynamic stability:	Well-conditioned core muscles significantly contribute to dynamic stability, and therefore optimal joint alignment when standing and moving.
Multidimensional strength:	Multidimensional muscular strength facilitates being and feeling strong, regardless of body orientation.
Integration:	Muscle-focussed movements can assist the integration of fascial change by supporting structural stability as well as an inner sense of stability and groundedness.

3 Slings Training Principles

Differentiation:	For dynamic balance between core stabilising and movement-oriented muscles, core stabilisation and general muscle conditioning exercises are differentiated.
Core stability exercises:	The repertoire of core stabilisation exercises incorporates all body positions and movement dimensions.
Toning exercises:	The repertoire of three-dimensional muscle toning exercises is versatile, including unidirectional and multidirectional movements with different loads and rhythms.

3 BENEFITS OF MUSCLE COLLABORATION

1. Active contribution to dynamic stability

2. Functional strength

3. Inner sense of support and stability

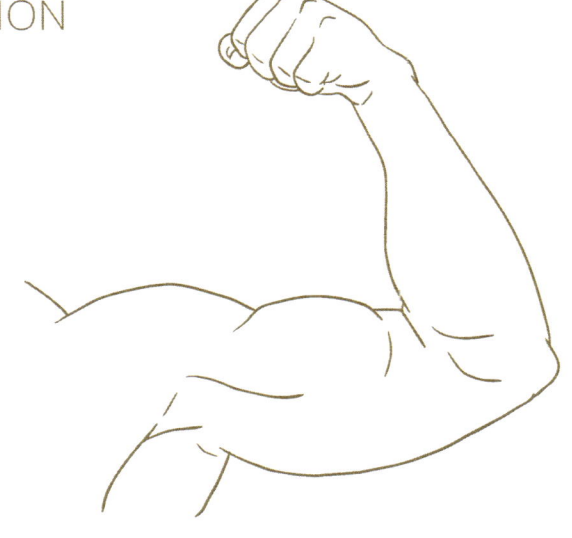

PART 2

The **force transmission** in fascia facilitates efficient communication among your body's systems.

3. FORCE TRANSMISSION

Fascia is a force transmission system. Through this mode of mechanical communication, it facilitates a bodywide response to local events, increasing movement efficiency, and decreasing strain on individual structures.

Let Me Ask You

How does tension in the neck make itself known in the soles of the feet? How can the big toe influence core stability all the way up to the head? How does strength in the right hip empower the left shoulder?

Fascial force transmission is one important part of the answer to all three questions.

By conveying changes in one part of the body to the rest, force transmission changes fascial tension distant to the source, increases movement efficiency, and decreases strain on individual parts.

A Good Reason To Care

You have good reason to get onto the mat and care for the communication skills of your fascia regularly if bodywide core stability and movement dexterity interest you.

LOCAL AND BODYWIDE MECHANICAL COMMUNICATION

Fascia is a multidimensional and active force transmission system, not – as assumed in the past – a unidirectional and passive connective tissue.

MECHANICAL MOVEMENT COORDINATION

Communication of tensional changes

Force transmission is a mode of mechanical communication. Changes in one part are communicated to the rest of the body, enabling bodywide responses to local events. This means that the whole body immediately responds to alignment changes in a well-orchestrated manner.

Bodywide response to local events

Say you walk along the street. By turning your head to check for oncoming traffic, the tensional changes in the fascia of the neck muscles are communicated down the back, the legs, and all the way to the feet. Fascially and neuromuscularly, your body acts as a coordinated whole, not as a head that is chauffeured around by a disconnected body. So far, so good. You keep walking and stumble on the sidewalk. As you catch yourself, the fascia in your feet and ankles has already sent messages around the body, tensioning fascia and triggering muscle actions to prevent you from falling. Force transmission in functional fascia is a step ahead of your conscious movement decision-making. Naturally, none of this happens in isolation, but in coordination with the nervous system.

Stretch Reflex
At this point, it is necessary to mention the stretch reflex. Fascial tension stimulates mechanoreceptors, which trigger the appropriate muscle contractions via the spinal cord. These contractions support prompt and coordinated responses to changed joint alignment. It is a reminder that all movement – deliberate or unintentional – is a neuro-myo-fascial synergy.

FORCE TRANSMISSION

Within and between myofascial structures

Force transmission occurs both within muscles and throughout the fascial system; however, there are significant differences in efficiency. Generally speaking, the more collagenous and architecturally organised a fascial structure is, the more efficiently it transfers force. As an example, collagen-rich, unidirectional tendons are excellent force transmitters.

Lucas and Karin transmitting good energy in series at the central station in Berne in Switzerland

In contrast, the watery loose fascia in which fibres crisscross is much better suited as a sliding medium than a force transmitter, for functionally good reasons.

High collagen content facilitates force transmission

While this general statement holds important truths, keep an open mind. There are no rules without exceptions – and at the moment, the rules are not even fully established. We are still in the early days of understanding fascia. Take highly collagenous retinacula as an example. Until recently, these structures were believed to primarily stabilise tendons and assure optimal force transmission. Research now indicates that the proprioceptive function of a retinaculum may outweigh its retainer role.

In Series and Parallel

Force is transmitted between muscles via their associated fascia, both in series and parallel.

In series: Force is transmitted consecutively from muscle fibres, to muscle fascia, to tendon, to ligament, to periosteum, and from there to the adjoining ligament, tendon, muscle fascia, and so forth.

In parallel: Force transmission occurs in laterally adjacent fascial structures; from the superficial fascia to the periosteum, the periosteum to the superficial fascia, and between neighbouring muscles.
For example: from muscle fibres to endomysium, to perimysium, to epimysium, and from there to the adjoining aponeurotic and superficial fascia.

FORCE DISTRIBUTION AND CONDUCTION

Fascial force transmission includes both the multidimensional distribution of strain and the unidirectional conduction of energy. This means that it decreases stress on individual structures and enhances performance. For conceptual clarity, let's look at these two distinct roles of fascia.

Fascial System: Force Distribution System

Unloading individual structures by strain distribution

The fascial system is very adept at multidimensional force distribution. By widely dispersing strain, less load is placed on individual myofascial structures and joints. Seen from this perspective, we recognise the fascial system as a force and strain distributor.

Myofascial Meridians and/or Slings: Force Conductors

Utilising myofascial meridians as force conductors

When considering the role of the fascia in mechanical communication, we focus on in-series force conduction between muscles, which brings us to the myofascial meridians of the Anatomy Trains concept created by the brilliant Thomas Myers.

Facilitating efficiency in everyday and athletic activities

Originally developed as a teaching tool, Tom identified and named twelve myofascial meridians throughout the body that he called Anatomy Trains. In essence, myofascial meridians are continuities of muscles and fascia that conduct force over long distances, enhancing movement efficiency in everyday as well as athletic activities. Think of throwing a ball with just your arm and shoulder. Then imagine a windup, and the power you gain from the energy contribution of the resilient myofascial sling that connects your arm with the opposite leg. Clearly, performance is increased! Additionally, the functional force conduction allows each myofascial unit within a sling to be less taxed, preventing overuse injuries.

This model demonstrates that moving bodies don't – and actually can't – operate one muscle at a time in isolation. Neither can a myofascial meridian be engaged separately. While it makes sense to focus on one or two of the Anatomy Trains lines in practice, it is important to remember that force is also transmitted between myofascial meridians and between other myofascial continuities.

Tom and Karin utilising their myofascial meridians in connection with one another by the river Sense in Berne in Switzerland

FUNCTIONAL FORCE TRANSMISSION

Tensile strength, fibre organisation, glide

Naturally, there is a difference between the possibility of efficient force transmission and its actual occurrence. Prerequisites for functional force conduction are:

- Sufficient tensile strength in the force transmitting fascia
- A well-organised fibre architecture in the force transmitting fascia
- Adequate glide between the force transmitting fascial layers

Insufficient and Excessive Force Transmission

Disorganised, adhered

Optimal force transmission is compromised – either insufficient or excessive – when fascia is disorganised (felty) or adhered. In other words, when there is a lack of structural organisation and inter-fascial glide.

Communication is disrupted, therefore ineffective

Insufficient force transmission means that sufficient tensioning is restricted. Imagine you are holding an exercise band in your two hands. When pulling on one end, the whole band tensions. You feel the pull in your other hand and a change in myofascial activity up into your arm and shoulder. Now imagine that someone is holding the middle of the band tightly while you are pulling on one end. The band tensions from your pulling hand to the other person's hand, but no further; the force transmission to your second hand is disrupted.

Communication is overdone, therefore restrictive

Excessive force transmission means that too much tension is transferred onto adjacent fascial layers and associated muscles. For example, when neighbouring fascial layers of 'antagonistically' working muscles are adhered to each other, the contraction in one muscle can create tension in the fascial envelope of the other, leading to movement restrictions. These restrictions might seem to stem from an inability of the antagonistically working muscle(s) to relax. However, the main issue is rigidity in the surrounding fascia, triggered by an excess amount of force that stems from adjoining myofascial units.

MOVEMENT MATTERS

Before moving on to the practical application, let's briefly visit fascial architecture 'Rule Number One': form follows function and function alters with form. When fascia is regularly, sufficiently, and multidimensionally loaded, it works well as a force transmission system. When fascia works well as a force transmission system, its architecture remodels accordingly, which improves movement efficiency. It is a positive self-sustaining process. So, dear reader, keep moving, the functionality of your fascia relies on it!

Form follows function; function alters with form

Healthy Loading

As a force conductor, fascia is strengthened by loading it through movement. Avoiding movement weakens fascia. 'Use it or lose it' applies. All movements within healthy parameters are beneficial. However, if you are after stronger fascia that transmits force more effectively, load intensity matters. It is suggested that high load in short intervals stimulates collagen production more efficiently than low load exercises. Because fascia remodels slowly, it is important to note that a strategy of gradually increasing load should be implemented to prevent injury.

Use it or lose it

Intensity aside, healthy loading also refers to dimensionality. Because the fascial system is multidirectional in nature, maintaining or gradually incorporating three-dimensional movements at different speeds is vital to assure healthy force conduction and strain distribution in the fascial system.

Multidimensional loading within healthy parameters

PART 2

SLINGS IN MOTION

ANATOMY TRAINS IN MOTION

The fascial system in general and the myofascial meridians specifically are engaged as force transmitters in every movement we make. The question is not if, but how can we facilitate force conduction and distribution consciously when training?

Focus on one or two myofascial meridians

In Slings, we pinpoint our movement focus on one or two myofascial meridians, thus clarifying our objectives and encouraging us to establish clear communication in our teaching. Within the short-term scope of a lesson and over a longer-term training plan, the complexity of the movement sequences and the number of myofascial meridians in focus increases. Additionally, the focus switches faster between the different lines.

DELIBERATE MOVEMENT STRATEGIES

For clear movement intention, we differentiate exercises that emphasise more linear force conduction or more widespread force distribution.

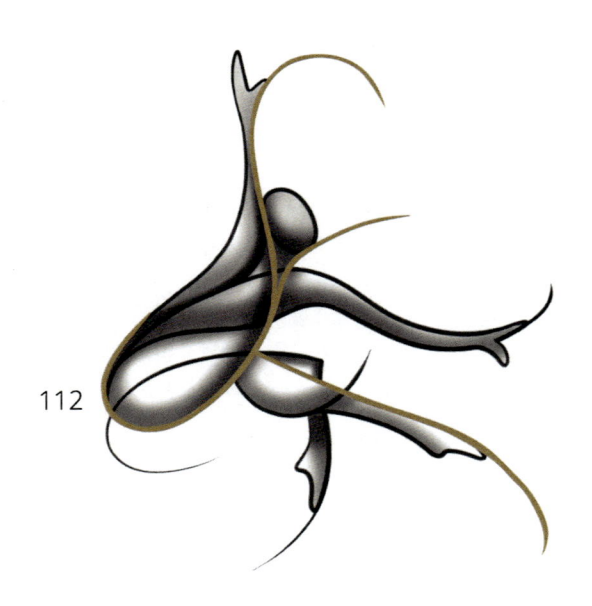

Clear and Unidirectional

Exercises that promote in-series force transmission generally have clear lines of tension and a certain sharpness that is perceptible. Frequently a myofascial sling is tensioned in a specific direction. However, the direction can change within a movement sequence. The rhythms range from rather slow to dynamic. Frequently used are:

- Active lengthening exercises with clear lines of tension
- Muscle-focussed strengthening with clear lines of tension
- Elasticity-enhancing bouncing and swinging exercises

Clear and perceptible lines of tension

90/90 Lunge & Arm Circle

90/90 Lunge with Side Bend & Arm Circle

Generous and Multidimensional

Exercises that promote more widespread force distribution are generally further reaching, expansive, multidirectional, and performed at a relatively slow pace. Frequently used are:

- Multidimensional, tensegral motions
- Glide-enhancing movements

Tensegral and gliding motions

90/90 Side Stretch with Spiralling

Permeability-Enhancing

Hydrating and gliding motions

Both force conduction and strain distribution require permeability in the fascial system. Permeability refers to a healthy degree of tissue adaptability and glide. Frequently used are:

- Hydrating motions
- Glide-enhancing movements
- Self-massage exercises

Curl Up Butterfly & Twist with Back Massage

Back Massage

A Word About Bony Stations

Bony stations are places where a tendon blends into the periosteum of a bone. Through repetitive strain, bony stations can be overloaded, which can lead to excessive fibrosity in the associated fascia as well as pressure sensitivity or pain on and around the bony station itself. Force transmission over long distances includes tensional changes across bony stations. Therefore, it is worthwhile to pay attention to these frequently overlooked regions and deliberately attend to them during the practice.

Gentle Stimulation

When done regularly, self-massage with a soft massage prop on and around a bony station can be beneficial to enhance fascial permeability.

Glide All-Around

To unload and improve overall functionality in bony stations, we enhance glide between the myofascial layers around them.

Twisted Pelvic Tilt with Sacrum Massage

SLINGS MYOFASCIAL TRAINING TECHNIQUES AND AIMS

The following Slings Myofascial Training Techniques and Training Aims shown in gold, bold font are directly related to force transmission.

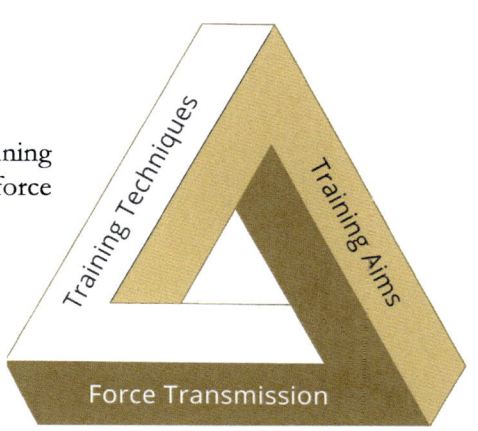

12 Myofascial Training Techniques
Techniques that optimise force transmission:

1. Stabilising
2. Toning
3. **Pushing, Pulling, Counter-traction**
4. **Active Lengthening and Expansion**
5. **Spiralling**
6. Hydrating
7. **Gliding**
8. **Domino Motion**
9. **Bouncing and Swinging**
10. Massaging
11. Active Ease
12. Melting and Invigorating

12 Training Aims
Four top achievements of force transmission:

1. **Dynamic Stability**
2. **Multidimensional Strength**
3. **Limberness**
4. Movement Rhythm
5. Elasticity
6. Tissue Nourishment
7. **Adaptability**
8. Resilience
9. Somatic Trust
10. Movement Courage
11. Mañana Competence
12. Kinaesthetic Intelligence

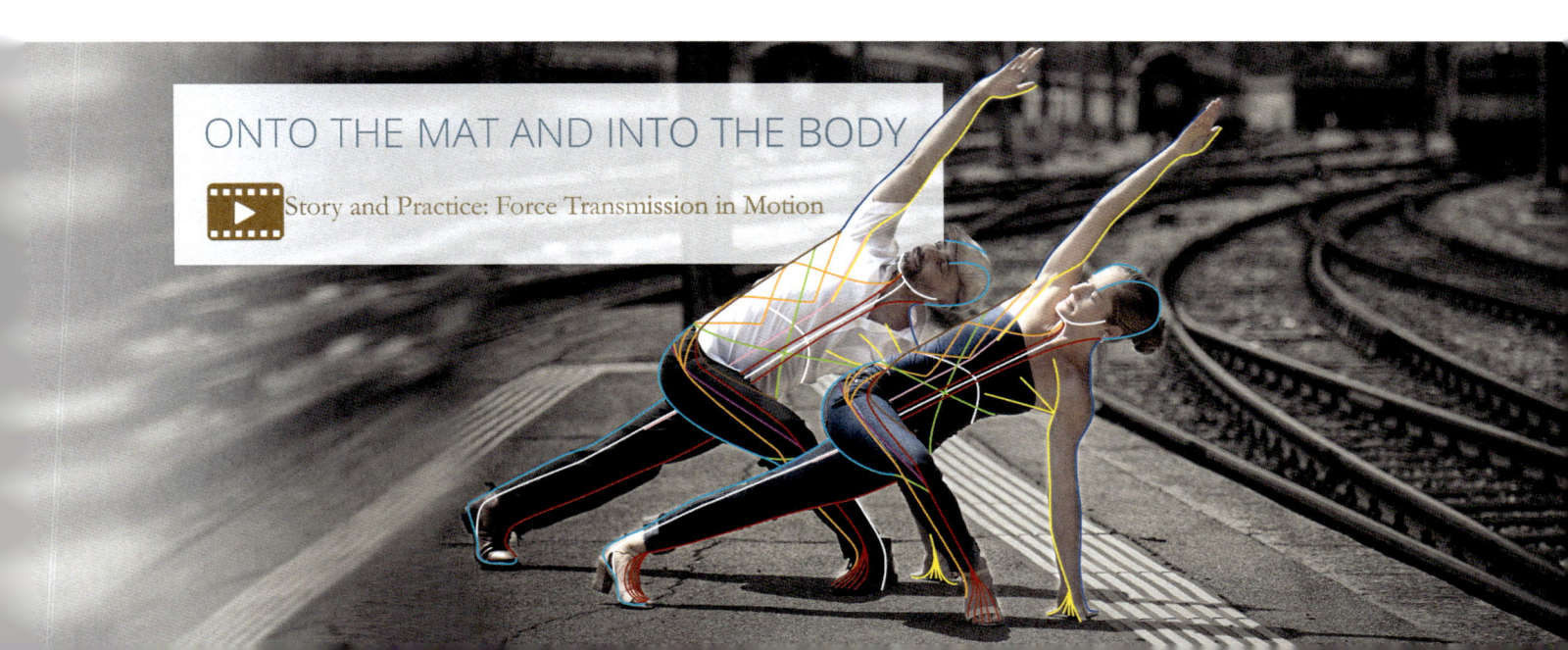

ONTO THE MAT AND INTO THE BODY

Story and Practice: Force Transmission in Motion

PART 2

IN A NUTSHELL
THINGS WORTH KNOWING ABOUT FASCIAL FORCE TRANSMISSION AND MOVEMENT

4 Things About Force Transmission

4-D force transmission:	Fascia is a multidimensional and active force transmission system.
Mechanical communication:	Force transmission is a mode of mechanical communication.
In series and parallel:	Fascia transmits force in series within a muscle and along myofascial slings, and in parallel to laterally adjacent myofascial structures.
Bodywide communication:	By communicating changes from one area to the rest of the body, fascia enables a bodywide response to local events.

4 Things About Force Distribution and Force Conduction

Distribution and conduction:	Fascial force transmission includes both the multidimensional distribution of strain and the unidirectional conduction of energy.
Conceptual differentiation:	For clear training intention, strain distribution and force conduction are differentiated.
Strain distribution:	Force distribution disperses strain widely, decreasing strain on individual structures, therefore supporting health and longevity.
Force conduction:	Force conduction facilitates mechanical, fascial communication over long distances, which increases movement efficiency.

4 Things About Functional and Dysfunctional Force Transmission

Functional transmission:	Prerequisites for functional force transmission are sufficient tensile strength and a well-organised fascial architecture in combination with sufficient inter-fascial glide.
Structural dysfunction:	Disorganised (felty) and adhered fascia inhibits healthy force transmission.
Compromised transmission:	Dysfunctional fascia can lead to inefficient or excessive force transmission, which decreases movement efficiency and ease.
Healthy loading:	Fascia needs to be regularly and multidimensionally loaded within healthy parameters.

4 Slings Training Principles

Myofascial meridians:	The movement focus is often on one or two myofascial meridians only, to maintain a clear training intention.
Clear and sharp:	Generally, exercises promoting force conduction have clear lines of pull.
Generous, multidimensional:	Generally, exercises promoting force distribution are generous and multidimensional.
Hydrating and gliding:	Generally, exercises promoting fascial permeability facilitate fluid flow and glide.

3 BENEFITS OF FASCIAL FORCE TRANSMISSION

1. Local and bodywide communication between muscles and fascia

2. Movement efficiency

3. Unloading of individual structures

PART 2

The **adaptability** of fascia makes you and keeps you adaptable for a lifetime.

4. ADAPTABILITY

Fascia is remarkably adaptable and remains so for a lifetime. Its adaptability enables gradual remodelling of postural and movement patterns to optimally support the body, as well as immediate somatic responsiveness when moving in known and unexpected ways.

Let Me Ask You

Why do we have a tough iliotibial band on the outside of the thighs? Why does a giraffe have a long ligament that acts like a giant rubber band from the back of the skull all the way down to the base of the tail? Because fascia models according to how humans and giraffes function. We are bipedal and therefore need extra-strong hip support. Giraffes have a very long, extremely mobile neck and therefore need extra-strong spinal support.

Not only does fascial architecture shape to the way we stand, move, think, and feel, it can also give us some 'giraffe qualities': physical responsiveness, agility, and resilience.

A Good Reason To Care

If postural ease matters to you and you wish to take movement hurdles in stride, or even turn them into an advantage, you have good reason to get onto the mat and regularly care for your fascial adaptability.

POSTURAL ADAPTATION AND MOVEMENT ADAPTABILITY

Lifelong adaptability

Fascia is remarkable in its ability to alter its own organisation. Over time, fascial architecture changes to meet postural and movement demands. In a more immediate context, fascia can and should alter its state to adapt to changing body alignments. The best news is that this quality is available for a lifetime, and it can be deliberately influenced. Note to self – there is always something you can do to reshape your inner architecture for the better!

Something can be done to feel better

Long-Term Postural Adaptation

Reinforcement of patterns; conscious or not, for better or worse

By remodelling the structural architecture unhurriedly, fascial adaptability reinforces our habitual postural and movement patterns; conscious or not, for better or worse.

Immediate Movement Adaptability

Facilitation of physical responsiveness

The natural resilience of fascia permits immediate movement adaptability. This is true for all motion; reflex-based movements like a startle response, unconscious activities such as turning in your sleep, deliberate actions when doing yoga, and unintended events, like stumbling on the sidewalk. Whenever you move, your fascia adapts to support you, or so it should; ideally, tout de suite!

Adrian moving and adapting easily on the Esplanade in Fremantle in Western Australia

Karin adapts to the timeframe of fascia by the Novodevichy Convent in Moscow in Russia

The Difference is in Timing

Naturally, the degree and speed of long-term adaptation and short-term adaptability vary by body part and person. As a general rule, the looser and more fluid the tissue is, the more readily it changes. Fascia containing less water is typically denser and more collagenous. Therefore it is remodelling at a slower pace.

The key difference between architectural adaptation and movement adaptability is the timeframe in which change occurs. Architectural adaptations in the fascial fabric occur slowly over months, while movement adaptability is an instantaneous response to changing body alignments.

Postural adaptation is slow; movement adaptability is immediate

Fibroblast cells remodel fibrous patterns

Cellular Activity

The main cells in charge of remodelling the fibrous component of fascia are the fibroblasts. By laying down and breaking down collagen, the activity of these cells slowly shapes the fibrous architecture, making the tissue firmer or more pliable. Think of a fascial pattern as a structural reinforcement. Ideally, it supports your postural alignment and movements in such a way that demands placed on the body are easily handled. As always, ideal is dynamic balance within a healthy spectrum and not a certain state to which the body clings. At either end of the spectrum there is a gradual tipping place where adaptability or lack thereof becomes limiting, unsupportive, or plain unhealthy. Exaggerated collagen density creates stiffness, leading to decreased physical adaptability and its subsequent domino effect of negative consequences. Collagen density that is too low for the loads placed on the body leads to instability, therefore structural disintegration (yes, you can be fascially too adaptable for your own good!). Both stiffness and instability eventually alter postural patterns, which in turn changes the way the body adapts – or not – at any given moment.

Fasciacyte cells facilitate ground substance pliability

Fascial adaptability does not only depend on the organisation of the fibrous network, though. It also alters with the degree of viscosity, therefore fluidity or density of the tissue. This brings us to the lubricating fasciacytes. Activity in these cells aids fascial fluidity by producing water-absorbing hyaluronan, which supports the pliability of the ground substance. That said, these hyaluronan molecules can become excessively 'bound', for example, due to lack of movement. This can lead to such a high degree of viscosity that the ground substance becomes a restricting glue rather than a lubricating gel.

LONG-TERM POSTURAL ADAPTATION

Slow remodelling according to use

Let's look at the development of deeply imprinted fascial patterns and gradual, functional remodelling more closely. Essentially, fascia remodels according to use. When it is repeatedly engaged in the same postural or movement patterns, little by little, fascia rebuilds its collagenous architecture to match the demands placed on the tissue. The architectural adaptations include changes in tissue composition and fibre alignment.

Tissue composition: The ratio of fibres and watery ground substance

Fibre alignment: The organisation of collagen fibres

Feedback Loop: Function and Form

"The dancer and the dance: without one, you have neither."   Mina Bissell

Vasily dancing in front of the Bolshoi Theatre in Moscow in Russia

Fascial (re)modelling is not a one-way street. It is a circular process during which form follows function and function alters with form. This means function and form create a feedback loop where one changes the other.

Form follows function:	Fascial architecture (form) develops according to use (function).
Function alters with form:	How fascia is used (function) changes with its architecture (form).
Feedback loop:	Altered use leads to different architecture and altered architecture leads to different use.

Practically Speaking

Your fascial patterns shape to the way you move and the way you move changes with the shape of your fascial patterns. Take the iliotibial band as an example:

Form follows function:	The well-structured, close-knit collagen organisation (form) of the iliotibial band developed because of upright standing and walking (function).
Function alters with form:	Lateral pelvic stability and walking efficiency (function) change with the collagen organisation (form) in the iliotibial band.

Functional Self-Regulation

Fascia regulates itself

If fascial remodelling is an ongoing process, you might ask yourself, what system commands it. For example, you might wonder at what point does the iliotibial band stop getting thicker? A valid question, because after all, some people load it for a century or more. The answer is, the fascial system regulates itself. In that sense, fascia has its own intelligence and gets tougher or more pliable based on the demands placed on the body to optimally support it. Once the iliotibial band reaches a collagenous thickness that is supportive of a person's activities, collagen production and breakdown balance each other to maintain the current fascial architecture.

To have a basic understanding of how fascial self-regulation works, let's look at the mechanical aspect in a very simple and linear way. Keep in mind; it is never as simple as a model might suggest. Our physical reality has many variables.

Collagen is produced according to demand

How It Mechanically Works

The building-up process:
- The increased load on fascia stimulates fibroblast activity.
- The cells lay down more collagen.
- As a result, the tissue toughness, and therefore stability, increases.

Maintenance mode:
- Once the tissue has reached the degree of stability that matches the loading, the fascial architecture meets the functional demands.
- Because of the increased stability, the tissue deforms less easily, which decreases the stimulation of cells and their subsequent collagen production.
- The activity of the cells settles in maintenance mode. The tissue retains its current degree of stability.

The Balancing Act of Holding On and Letting Go

Dynamic equilibrium in the fascial system is a lifelong balancing act.

By standing and moving, you develop strong and supportive fascial straps, sheets, and envelopes throughout the body. Their close-knit collagenous architecture provides extra stability and durability, empowering you to function well, in the way you need and want. Nevertheless, your body is meant to move freely in all directions, reflected in the multidirectional organisation and watery nature of fascia. For movement ease, fascial stability and fascial adaptability are equally important. The interesting and sometimes challenging task is to dynamically balance these two qualities throughout the natural fluctuations of life.

Reason to Change?

Fascia adapts for a reason; conscious or not, by choice or accident, for better or worse.

There is no question that fascia adapts for a reason. The question is: are the adaptations helpful or not (anymore)?

Fascia always adapts for a reason

Redefine Your Fascial Body Map
Like the brain, fascia remains malleable throughout a lifetime – at least to a certain degree. Naturally, you can't reverse time, but, starting today, you can create a refined version of your fascial body map. Your fascia is ready to change when you are.

Fascia changes when you do

Fascial Patterns Are Complex
Our fascial system is full of patterns created for countless interconnected, unconscious, and deliberate reasons. In truth, fascial organisation is highly complex. Consider a particular kind of pattern, the kind we actively create and reinforce against our better judgement. We sometimes intentionally engage in activities that unintentionally hurt us. When becoming aware of this, the question is: at what point do the negative consequences of a myofascial pattern become intolerable, and change becomes necessary? The most common answers to this question include diminished life quality, the inability to perform activities as needed or wanted, and of course, the presence of pain. All of them are certainly strong motivators to unwind antagonistic patterns, so why might we not act accordingly?

PART 2

Do a little thought experiment: think of something you keep doing that obstructs your physical wellbeing. It could be running, although it triggers knee pain, stretching despite the pelvic discomfort it creates, wearing shoes that hurt the toes, or sitting too much, undeterred by the increasing backache. Got it? Now think of why you still do it. That's right, every pattern has its reasons, and sometimes they outweigh – or in the short-term seem to outweigh – the negative consequences. Many of us engage in activities or disengage from activity to feel temporarily better in one way or another, regardless of the ensuing long-term consequences. In that sense, the immediate benefit the activity affords us outweighs the pain it creates. This, of course, is not the end of the story; it is only the first layer, which lands us amid the complexity of humanness.

If philosopher René Descartes had been right in his proposition that the mind governs the body from a safe distance, we wouldn't have this human dilemma. Instead, we would simply stop or never start with activities that harm us. As brilliant a thinker as he was, Descartes wasn't right. We are thinking and feeling physical beings who create structural patterns that reflect our complex inner nature.

Unwinding patterns is multilayered – it includes letting go

It is good to remember that unwinding myofascial patterns is a somato-psycho-emotionally multilayered process, which includes giving up something that – knowingly or not – held value or a least gave us something at one point.

I feel, and I think; therefore, it's complicated

It is Complicated
Descartes famously said, "cogito ergo sum", commonly translated into "I think, therefore I am" and less commonly into, "I doubt, therefore I am". Let's embrace the latter. You don't have a mindless body or disembodied mind – instead, you are your body and your mind. From here onward, let's work with and therefore acknowledge that our structural patterns are a part of who we are and a part of being human. It is indeed complicated.

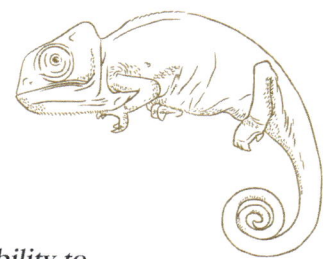

IMMEDIATE MOVEMENT ADAPTABILITY

Movement adaptability refers to the fascia's ability to temporarily alter its structural patterns and consistency when the body is moving. It is the kind of conformability you want from a snug pair of jeans. It fits well (stability), yet the fabric is sufficiently stretchy (adaptability) to let you move the way you want (movement freedom).

Fascia that adapts readily makes your body immediately responsive to alignment changes. Say, you stumble on the pavement or lift a heavy shopping bag out of the boot of your car while being distracted by a talkative neighbour. You want your fascial system to adapt effortlessly and support these unpractised activities to prevent injury.

Because what we practise on the mat needs to benefit the other 160-something hours of the week, we deliberately prepare the body to function more efficiently in known and expected ways.

Ability to readily adapt to changing body alignments

Being prepared for the unexpected

Fascial Resilience

If the key phrase for long-term adaptation is "form follows function and function alters with form", the word to remember for immediate fascial adaptability is resilience. The inherent tensile strength and elasticity of fascia make it resilient, an essential feature of immediate adaptability. Sometimes movement adaptability is equated with pliability. However, in this context, sufficient pliability is only fifty percent of the equation. The other fifty percent is the tissue's ability to recoil elastically to its original shape.

Tissue resilience supports dynamic stability and optimal force transmission throughout the fascial tensegrity. It also modulates muscular activity. All of it enhances your physical responsiveness and movement agility.

Tissue resilience supports immediate movement adaptability

Limiting Factors and (Re)Training Resilience

Limiting factors are – as always – manifold and correlated. For example, disorganised felty collagen, tissue adhesions, gluey ground substance, and excessive rigidity all negatively affect fascial resilience. The same is true for locked long (chronically lengthened) fascia that has lost its fluidity and elastic collagen architecture (spirals and crimp).

The good news is that fascia remains pliable, and resilience can be (re)trained with versatile movements in which direction, load, and rhythms are varied.

PART 2

SLINGS IN MOTION

You were not born with your current patterns, nor did you inherit them; instead, you diligently — and for legitimate reasons — created them. Your present fascial patterns are a malleable map reflecting your movement history as well as your movement potential. With awareness and perseverance, you can create a refined version at any time — if you choose.

REPATTERNING FASCIAL ADAPTATIONS

Long-term aims are postural ease and optimal movement functionality

A long-term goal of Slings Myofascial Training is a dynamically balanced body that has a comfortable degree of postural ease and movement freedom. This requires consistent fascial repatterning to either unwind patterns that are no longer helpful or create new patterns that support the changing demands placed on the body. As teachers, we don't demand change from our clients. Instead, we facilitate change with customised, posture-based training programs in a one-on-one setting or more generally in a group format, which is still amazingly effective.

Short and Long-term Lesson Planning

Movement adaptability is an immediate training aim with long-term benefits, while postural repatterning is a long-term goal with regular, immediate benefits.

Adaptability: Fascial and, therefore, movement adaptability can be progressively trained in every lesson.

Adaptation: To change fascial adaptations, a far-sighted training approach must be taken. Because fascial architecture renews slowly, a timeframe of months is needed for sustainable structural repatterning. Still, the joys and somatic benefits that come with the process are motivating and worth appreciating.

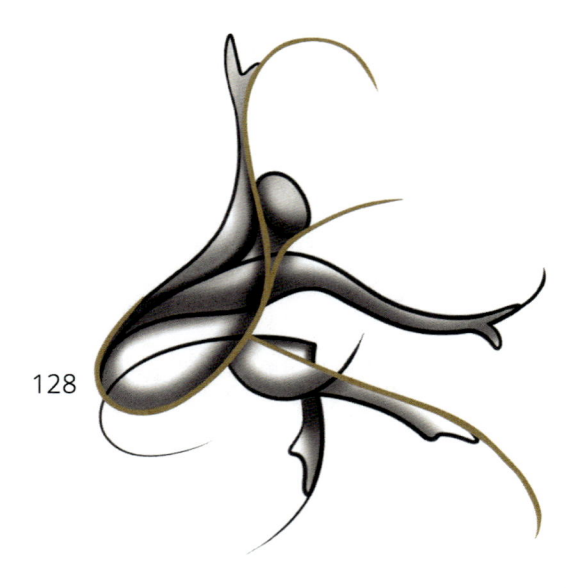

Take and Give

Both in group and personalised lessons, we recognise that each body has a meaningful story. Unwinding patterns requires letting go of something that played a long-standing role in an individual's existence. For this reason, we combine movements designed to unwind patterns with rebalancing and stabilising exercises. In that sense, we take a bit and give back a bit. If myofascial patterns are changed too quickly or without the necessary somatic support, it might lead to structural chaos rather than the structural integration.

Unwinding in combination with rebalancing and stabilising

In and Out

Sometimes on purpose and sometimes by chance, a person begins by moving deeper into an existing pattern. This may sound counter-intuitive, but this process can actually help unwind that very same pattern. Imagine a string with a knot. How do you untie the knot? Do you pull firmly on either end of the string? Of course not; that would tighten the knot. Instead, you push the strings on either side of the knot closer together to soften the fabric, making it easier to undo.

Moving in to move out of a pattern might be necessary

Somatic patterns can behave similarly. By opposing them, they might resist and tighten even further. By moving into the pattern in a relaxed manner first, the fascial tissue may soften, and the subconscious may understand that there actually is a pattern to unwind – both of which can assist the rebalancing process. In this practice, movements are always done on both sides and rebalanced with bilateral exercises. Therefore, no mover gets or stays stuck in the patterns they might have moved into more deeply.

PRACTISING MOVEMENT ADAPTABILITY

Adaptability is practised in the play zone and challenge zone

Fascial adaptability can be (re)trained with versatile exercises that utilise all of the fascial qualities. This means that movement directions, loading intensities, and rhythms are varied intentionally and sensibly.

To become more adaptable, we need to move out of our 'comfort zone' and into the 'play zone' and 'challenge zone' regularly. For this reason, contrasting exercise sequences and 'unusual' moves are deliberately incorporated into the practice. By repeatedly moving back into the comfort zone, we calm the system, rebalance the body, and facilitate the integration of change.

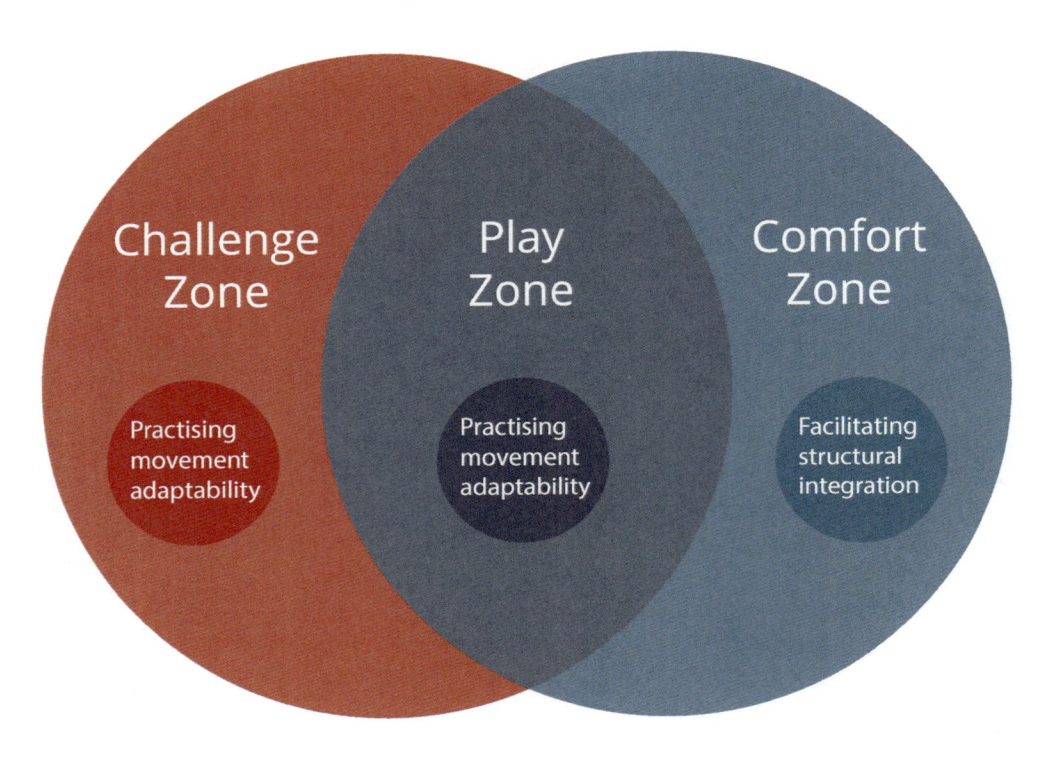

Integration is facilitated in the comfort zone

Challenge Zone

Contrasting sequences and testing exercises

The challenge zone is home of movements that test our kinaesthetic abilities, which includes physical skills and emotional awareness.

We purposefully blend exercises with different intensities and varying rhythms in contrasting movement sequences, weaving extra challenge in here and there, adding to the excitement. The extra challenge can come in the form of a surprising rhythm change, a balance or coordination challenge, an endurance demand, or even a time of stillness – anything that constitutes outside the comfort zone.

Expansion of movement spectrum

Warrior II   Extended Triangle & Dynamic Warrior II

Dynamic Warrior II into Full Moon   Standing Split

Forward Fold Leg Stretch & Rolling Up   Arm Arc & Dynamic Plié with Arm Circle   Relevé with Arm Arc

Balancing Grand Plié   Crouching   Forward Fold Leg Stretch & Rolling Up

PART 2

Unusual and fun movements

Joy and preparation for the unexpected

Play Zone

What are 'unusual' and fun movements? That depends on how you define usual and what is fun for you, doesn't it? For simplicity's sake, let's just say usual movement is the kind of activity you do habitually, in your preferred training modality when everything goes according to plan. In that sense, unusual movement means unpractised or unexpected motion. I will leave it up to you to decide what's fun.

In the Slings play zone, we add angles and spirals to otherwise linear movements, turn and fold in strange ways, play with rhythms, and mess a bit with coordination. Here is where functional fun prepares you for the unexpected.

90/90 Dynamic Shift into Gate Pose

Rolling Like a Ball into Standing   Forward Fold Leg Stretch Rolling Up

Comfort Zone

What is the comfort zone? Or perhaps the more accurate question is when are you in it? You are in your comfort zone when you engage in movements that you have embodied. You can execute these movements with unconscious competence and ease. Naturally, the comfort zone is a subjective experience. While some people feel in tune with slow and mindful motions, others are at home when performing intense exercises. In your practice or when you are teaching, consider these different preferences. In the context of Slings, the comfort zone is the place where somatic change is integrated and rebalanced. Therefore, we recommend engaging mindfully in relatively simple movements that bring about a sense of inner balance and stability.

Embodied exercises

Integration of change and centredness

Bilateral balance:	Unilateral exercises are balanced and complemented with bilateral motions.
Stability and integration:	Simple core stability exercises are regularly incorporated to re-stabilise the body and assist integration of change.
Grounding:	Focussing on gentle engagement and toning of muscles can have a beautifully grounding effect. Hence, it is beneficial to complement the contrasting and unusual motions of the challenge zone and play zone with muscle-focussed exercises that bring you back to your centredness.

Leg Float Up

Pelvic Curl & Arm Arc

ONTO THE MAT AND INTO THE BODY

▶ Story and Practice: Adaptability in Motion

SLINGS MYOFASCIAL TRAINING TECHNIQUES AND AIMS

The following Slings Myofascial Training Techniques and Training Aims shown in gold, bold font are directly related to adaptable fascia and movement.

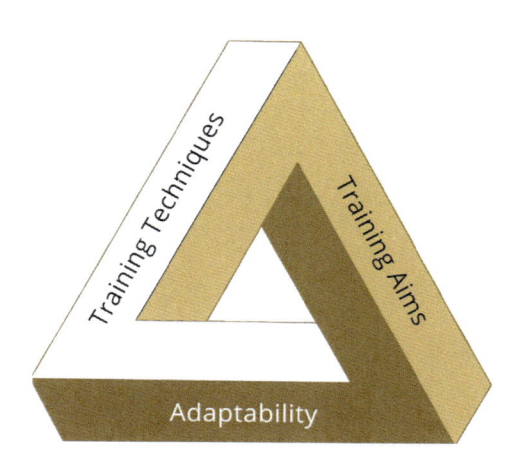

12 Myofascial Training Techniques
Techniques that optimise fascial adaptability:

1. Stabilising
2. Toning
3. Pushing, Pulling, Counter-traction
4. **Active Lengthening and Expansion**
5. **Spiralling**
6. **Hydrating**
7. **Gliding**
8. **Domino Motion**
9. **Bounce and Swinging**
10. **Massaging**
11. **Active Ease**
12. **Melting and Invigorating**

12 Training Aims
Four top achievements of adaptable fascia:

1. Dynamic Stability
2. Multidimensional Strength
3. **Limberness**
4. Movement Rhythm
5. Elasticity
6. Tissue Nourishment
7. **Adaptability**
8. **Resilience**
9. Somatic Trust
10. Movement Courage
11. Mañana Competence
12. **Kinaesthetic Intelligence**

IN A NUTSHELL
THINGS WORTH KNOWING ABOUT FASCIAL ADAPTABILITY AND MOVEMENT

4 Things About Long-term Fascial Adaptation

Lifelong adaptability:	Fascia is remarkably adaptable and remains, to varying degrees, malleable for a lifetime.
Adaptation meets demand:	Fascial architecture remodels to meet the demands placed on the body.
Feedback loop:	Body use and fascial remodelling create a feedback loop in which 'form follows function and function alters with form'.
Slow remodelling:	Fascial patterns develop gradually and remodel slowly over the timespan of months.

4 Things About Immediate Movement Adaptability

Movement adaptability:	Fascia temporarily alters its structural patterns and consistency during motion, thereby supporting immediate movement adaptability.
Tissue resilience:	Tensile strength and elasticity are characteristics of readily adaptable fascia.
Limiting factors:	Disorganised felty collagen, adhesions, or dense gluey ground substance are factors that limit fascial adaptability.
(Re)training adaptability:	Fascial adaptability is (re)trained with versatile exercises that utilise all Fascial Movement Qualities.

4 Slings Training Principles

Dynamic body balance: The deliberate unwinding of fascial patterns is done slowly and in combination with integration exercises.

Challenge zone: The 'challenge zone' contains contrasting sequences and testing exercises that expand the movement spectrum.

Play zone: The 'play zone' contains that bring joy and preparedness for the unexpected.

Comfort zone: The 'comfort zone' contains the integration of change and centredness.

3 BENEFITS OF FASCIAL ADAPTABILITY

1. Long-term postural and movement support
2. Immediate physical responsiveness
3. Lifelong potential to change for the better

PART 2

The **multidimensionality** of fascia enables you to move freely in all directions and rhythms.

5. MULTIDIMENSIONALITY

Fascia is multidimensional by personal design. It enables and supports the body's naturally diverse movement spectrum and, therefore, the freedom to move as wanted or needed.

Let Me Ask You

What enables you to move freely in all directions? What qualities do people still living in natural environments retain that many urbanised people gradually lose? The multidimensionality of fascia and movement diversity.

Movement is inherently multidimensional, just like the body. This is true for everyday activities such as walking and making the bed as well as for holistic movement practices and athletic training. Fascial architecture naturally reflects this trait, because form follows function. Fascia also responds to movement rhythm and is an important contributor to rhythmicality, which adds another dimension to the three dimensions of movement: time.

A Good Reason To Care

Everyday functionality, improved training performance, or simply having fun moving in all directions without restriction, are good reasons to get onto the mat and take care of the multidimensionality of your fascia regularly.

FASCIAL ARCHITECTURE ACCORDING TO MOVEMENT DIVERSITY

Before moving forward, let me explain the choice of the term multidimensional over three-dimensional.

Multidimensionality unites 3-dimensionality and rhythm

In regard to movement, three-dimensional commonly refers to motion in the three planes of movement: the sagittal plane, frontal plane, and transverse plane. For example, bending and arching the spine occurs in the sagittal plane, side bending in the frontal plane, and rotation in the transverse plane. When walking, the spine moves in all three planes. While it is helpful to differentiate planes of movement to understand basic mechanics, it is important to note that the indicated plane refers to the predominant and not the exclusive movement direction. When raising the arms sideways over the head, the predominant plane of movement is the frontal plane. However, while the arms move out and up, the shoulder joint also moves slightly in the sagittal and transverse planes. Hence, it is a unidirectional activity that is inherently three-dimensional.

And now, let's look at the fourth dimension: time. Time in this context refers to movement rhythm or the timing of muscle activation and fascial adaptability, which shows in the speed or slowness in which we move.

This concept of multidimensionality can be applied to fascia. The fibrous architecture of fascia is multidirectional (three-dimensional), and in combination with the ground substance, it adapts to particular stimuli in different rhythms (time).

NATURAL MULTIDIMENSIONALITY OF FASCIA

Movement diversity creates fascial multidimensionality; and vice versa

Shaped by our natural movement diversity, fascia is multidimensional by design. While the fascial system as a whole is multidimensional, the architectural patterns of individual structures vary greatly, ranging from highly organised to seemingly chaotic. Even though the collagen orientation of some fascial structures is predominantly unidirectional, functionally, they are still part of and act as members of a multidimensional tissue network. As an example, consider a tendon whose tightly-knit collagen is stringently organised.

By comparison, the fibres in the adjacent loose fascia are rather like crisscrossing cotton candy. Although their structural organisation is different, functionally, they are inseparably linked. While the tendon tensions unidirectionally, the surrounding tissue adapts in a non-linear fashion, which is important for the optimal functioning of the tendon.

Kurt, age 69, sustaining his multidimensional movement capacity in front of Parliament House in Berne in Switzerland

PART 2

At All Ages

When at the age of 98 James Wooden (former UCLA basketball player and head coach) was asked by Alan Castel about the keys to successful ageing, he said "love and balance - and besides that stay busy, stay active and have some variety."

Importance increases with age

The body naturally changes as we age. However, many of us lose our capacity for diverse movement faster than nature intended. Resulting movement limitations decrease our quality of life, and worse case, they disable us to do the things we need to or love to do. Our physical adaptability is also compromised, which increases the risk of pain and injury. The older we get, the more important it is to keep our movement spectrum as broad and as versatile as possible.

Kurt staying active and having movement variety in the old town of Berne in Switzerland

SLINGS IN MOTION

(RE)TRAINING MULTIDIMENSIONALITY

On playgrounds and Balinese rice fields, you can see wonderful movement diversity. While most of us start out well, many of us gradually lose the full spectrum of our multidimensional movement capacity as we grow up. We accept it as part of maturing and start to walk in straight lines to get from A to B efficiently. To conform to social norms, we straighten up our body, and when adult life becomes tiresome, we prioritise inactivity over activity. At one point, we might get hurt – or simply grow older. A well-meaning person advises us that twisting, bending, and moving freely should be avoided. They might add that this kind of movement is no longer healthy, has never been healthy, or even go so far to say that at this point in our life, it is outright dangerous. I will state my opinion, bluntly. The danger is not in the multidimensionality of movement; it is in the loss of it or the haste with which one attempts to regain it. When done with kinaesthetic awareness and patience, multidimensional movement freedom can be retrained in a safe and healthy way.

Multidimensionality can be retrained safely

Stuck in Line?
Useful as it may be for some purposes, by only training in straight lines, single planes, and limiting 'safe zones', we don't utilise and enhance one of our greatest fascial qualities: multidimensionality. According to fascia, your body is meant to move freely in all directions, at varying speeds, and with different intensities.

Linear training is insufficient

Twirling Spine
One additional word about the spine is necessary. Multidimensional spinal movement, or the restriction of it, is still a much-discussed and regularly debated topic in the lands of movement training and therapy. I believe that the spine and associated myofascial structures are built for three-dimensional movement and should be used accordingly. While different spinal regions have varying degrees of mobility, no area should be fixed or excluded from movement. On the contrary, the areas that are most restricted, underused, and fascially weakened or dehydrated need to be moved. Begin very gently. Give your back your undivided attention when reinvigorating the spine. Converting fixation into dynamic stability and immobility into movement freedom takes time. It is time well invested!

Multidimensional movement sustains spinal health

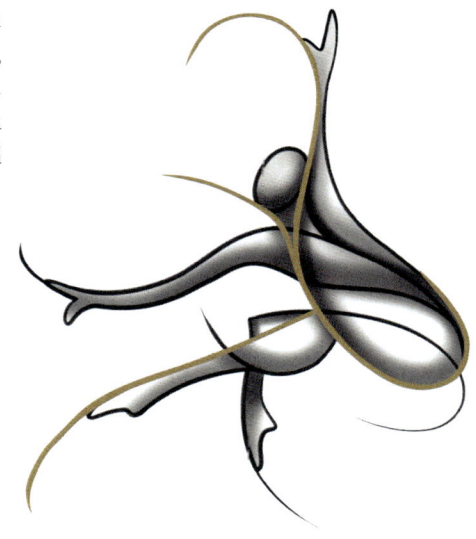

PART 2

Simone moving in diverse ways to maintain her everyday functionality in her parents' home in Môtier in Switzerland

Multidimensionality is (re)trained and sustained with versatile movement

Movement Diversity

Because "form follows function and function alters with form", the principle is simple: diverse movement trains and sustains the natural multidimensionality of fascia. In turn, the multidimensionality of fascia supports movement diversity. In practical terms, it means that the following is considered:

- All Fascial Movement Qualities
- Unidirectional and three-dimensional exercises
- Simpler and more complex exercises
- Varied intensities
- Changing rhythms

Training in the Timeframe of Fascia

Remember that fascia remodels slowly over the timeframe of months. Therefore, turning habitual linearity into multidimensional movement freedom requires patience and gentle persistence. To prevent unnecessary discomfort or worse, injury, progress thoughtfully in the rhythm of fascia, instead of the excitable speed of muscles, or the impatience of will.

Besides safety concerns, it is good to keep in mind that sustaining an adaptable, multidimensional fascial architecture is a lifelong project. There is no need or point in rushing a process that can't be rushed.

Retraining multidimensionality is a slow process

Progressive Training Strategy

A progressive training strategy needs some planning.

Baseline study: To start, take an honest look at where the body is now. Not the memory of where it used to be and not wishful thinking about where it could, or you think it should be.

Goal setting: Next, set clear and achievable short-term, mid-term, and long-term goals.

Stepping stones: Now, create stepping stones in the form of exercises that progressively work towards your goals. Start more linear and gradually add directional and rhythm changes. Of course, the movement spectrum increases with each stepping stone mastered, and in that sense, every step is a success.

Retraining multidimensionality is a progressive process

PART 2

MULTIDIMENSIONALITY TIMES THREE

Multidimensionality in motion and dynamic stabilisation

The concept of multidimensionality is applied to grand and subtle movements, as well as dynamic stabilisation.

Grand Multidimensional Movement

Multidimensionality is most recognisable and perceptible when it is expressed with generous movements of the spine, arms, and legs.

Foot Spring with Double Spiralling

Subtle Multidimensional Movement

Multidimensionality is more subtle when tensegral qualities such as ribcage expansion or spiralling movements of the shoulder and hip joints are added to larger, outwardly unidirectional movements.

Kneeling Triple Extension

Multidimensional Dynamic Stabilisation

The subtlest expression of multidimensionality is found in dynamic stabilisation exercises. Regardless if a stabilised body part is outwardly static or not, myofascial stability is always three-dimensional and includes variability in muscle tone.

Knee Lift

Single Leg Plié

Single Leg Pivot

Foot Spring

ONTO THE MAT AND INTO THE BODY

▶ Story and Practice: Multidimensionality in Motion

SLINGS MYOFASCIAL TRAINING TECHNIQUES AND AIMS

The following Slings Myofascial Training Techniques and Training Aims shown in gold, bold font are directly related to multidimensionality of fascia and movement.

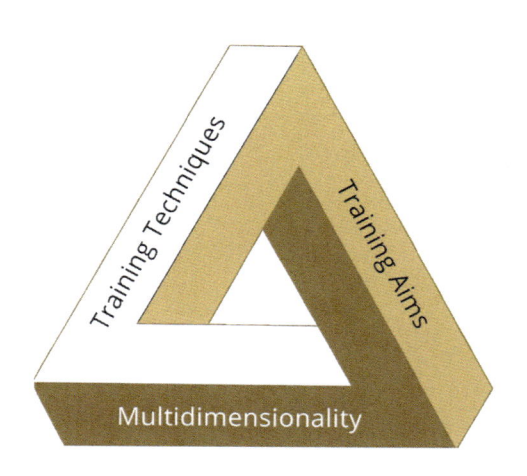

12 Myofascial Training Techniques

Techniques that optimise multidimensionality:

1. Stabilising
2. **Toning**
3. Pushing, Pulling, Counter-traction
4. Active Lengthening and **Expansion**
5. **Spiralling**
6. Hydrating
7. **Gliding**
8. **Domino Motion**
9. **Bouncing and Swinging**
10. Massaging
11. Active Ease
12. Melting and Invigorating

12 Training Aims

Four top achievements of multidimensionality:

1. Dynamic Stability
2. **Multidimensional Strength**
3. Limberness
4. **Movement Rhythm**
5. Elasticity
6. Tissue Nourishment
7. Adaptability
8. **Resilience**
9. Somatic Trust
10. **Movement Courage**
11. Mañana Competence
12. Kinaesthetic Intelligence

PART 2

IN A NUTSHELL
THINGS WORTH KNOWING ABOUT FASCIAL MULTIDIMENSIONALITY AND MOVEMENT

3 Things About Multidimensional Fascial Architecture

Time as 4th dimension:	Multidimensionality unites three-dimensionality and rhythm.
Form follows function:	Natural movement diversity creates fascial multidimensionality.
Multidimensional system:	Even though the architecture of some fascial structures is unidirectional, functionally, they are part of and act as members of a multidimensional tissue network.

3 Things About Multidimensional Movement

Natural multidimensionality:	The body is built to move freely in all directions, with various intensities, and at different speeds.
(Re)Training multidimensionality:	With awareness, multidimensionality can be retrained safely.
At all ages:	The older we get, the more important it is to keep the movement spectrum as broad and as versatile as possible.

3 Slings Training Principles

Movement diversity: Multidimensionality is intentionally (re)trained with versatile movements that consider all fascial qualities and vary directions, load, and rhythms.

Slow progression: Retraining multidimensionality is a slow process that is done in the rhythm of fascial change.

Gradual progression: The training is progressive, starting with differentiated and more linear exercises, gradually advancing to integrated, three-dimensional motions in different rhythms.

3 BENEFITS OF FASCIAL MULTIDIMENSIONALITY

1. Movement freedom in all directions
2. Adaptability
3. Resilience

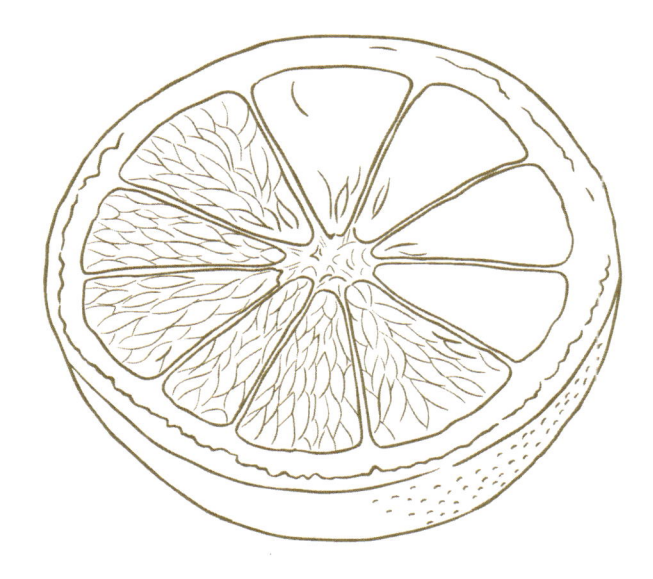

PART 2

The **fluidity** of fascia
makes them the vitalising ocean inside your body.

6. FLUIDITY

Fascia is a fluid-filled system and the watery habitat of cells. Fluid flow within this habitat is essential for health, vitality, and the body's self-healing capacity.

Let Me Ask You

Why do you feel healthier and more vitalised when you move regularly? Because you are!

The health of your cells relies on the health of their immediate living environment, namely the extracellular matrix. Fascia is extracellular matrix, therefore keeping your fascia fluid and well-nourished through movement strengthens your vitality and increases your energy, along with your body's ability to rejuvenate and heal. Compare the quality of fluid in a non-moving or stagnant water source to one in which there is fluid flow. When there is movement of the fluid, there is a means to nourish and cleanse the cells living within it.

A Good Reason To Care

You have about 70 trillion good, cellular reasons to get onto the mat and care for healthy fluid flow in your fascia.

VITALISING INNER OCEAN

"The concentration quotients of salts (NaCl, KCl, CaCl2) in interstitial fluid and in water of an ocean are nearly identical. Our cells are, in a manner of speaking, swimming gel-like structures in an ocean of interstitial fluids, and we are carrying that ocean around with us."
— Guido F. Meert

Fascia is the watery habitat of cells

Fascia is extracellular matrix (ECM), and extracellular matrix is the watery habitat of cells. Although the term extracellular matrix specifically refers to tissue around the cells, the broader definition of fascia also includes the cells that produce and maintain the matrix. In this context, most relevant are the fasciacytes, which seem primarily devoted to the production hyaluronan, a water-absorbing molecule that is essential for the optimal fluidity of the ground substance. Besides the water-repellent collagen and elastin fibres, the ground substance and water are the key ingredients of fascia. While some fascial water may be free-flowing, much of it is bound, like the water absorbed by a sponge. Considering that the concentration ratio of salts in fascial fluid and ocean water are similar, you literally carry around a vitalising, inner ocean.

Not All Seawater Is the Same

Fascial fluidity naturally varies

Not all seawater has the same salinity, and not all fascial structures have the same viscosity. The sliding, loose fascia is very fluid, while other more collagenous fascial structures, such as deep fascia, are much less watery. The state or fluidity of water itself also varies. It can form large regions of so-called structured water or liquid crystalline water, in which the molecules move closer together like a school of fish. The water remains fluid, yet it has increased viscosity (density).

Healthy Flow

The fascial system is infused with fluid channels. Like the fibrous network, the fluid channels are bodywide and continuous. Whatever the hydrodynamics or fluid ratio is in the tissue, fascia relies on sufficient fluid flow for optimal wellbeing and restoration. Nutrients, waste products, and messenger substances are transported by fascial water. Maintaining healthy flow cleanses and nourishes the tissue, aids rejuvenation, and speeds up healing.

Healthy Habitat, Healthy Inhabitants

"(Growth and malignant) cell behaviour is regulated at the level of tissue organisation, and tissue organisation is dependent on the ECM and the microenvironment."
Mina Bissell

Because fascia is the direct living environment of cells, it is a key influence on cell behaviour. A healthy fascial environment supports life-promoting cell behaviour, whereas a toxic and depleted environment increases the chances of disease. Naturally, many factors influence tissue health. Movement is not the only element, and it cannot cure everything, but it is a major contributor to physiological wellbeing and self-healing. Keeping your fascia moving and well-hydrated means taking care of your vitality.

Fluid flow in fascia is vitalising

Karin rehydrating her inner environment on one of the salt lakes in Western Australia

PART 2

SLINGS IN MOTION

"Get to know water properly and it will always be your trusted friend."
Sebastian Kneipp

FASCIAL HYDRATION

In practice, fascial hydration and vitalising fluid flow are facilitated with specific myofascial training techniques and sustained with the general diversity of the exercises. At this point, the question of fluid intake frequently arises.

Drinking is Not Enough

Drinking is not enough – fluid needs to be circulated and absorbed

As important as sufficient fluid intake is, fascial hydration requires more than drinking enough water. Fluid must be circulated through an immense network of capillaries and then dispersed and absorbed by the tissue. Movement is ideal for supporting this process.

Think of a dried-out garden and then a large amount of rainwater pouring down all at once; very little water seeps in. If the soil is looser, more and more fluid is absorbed and distributed. In Slings, we intend to create and maintain the optimal conditions for fluid to circulate and be distributed to all the 'nooks and crannies' of the fascial web.

Supplementing Fascial Nourishment Through Movement

Fluid flow is deliberately enhanced

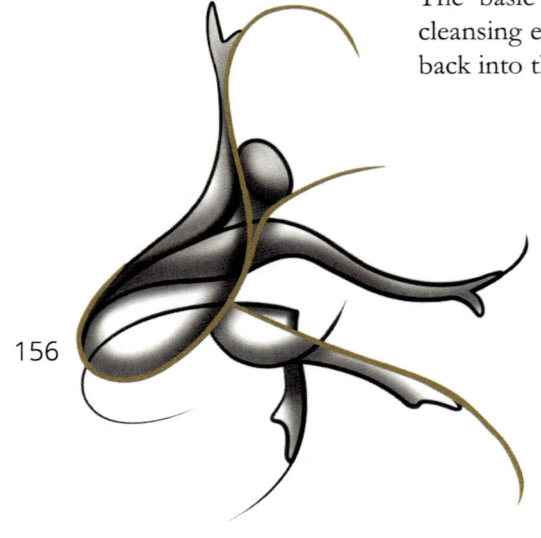

The key to fascial hydration is versatile movement. It is quite simple. Because modern life doesn't make staying in the flow easy, we promote fluid flow in fascia with specific training techniques and exercise sequences. It is a bit like taking a watering pot to assist nature. In light of climate change, this is not a bad idea for most of us!

The basic idea is to squeeze fluid out of the tissue, which has a cleansing effect. The consecutive release allows fresh fluid to be drawn back into the fascia to rehydrate, thereby nourishing it.

SPONGE EFFECT

The concept of squeezing and soaking a sponge is the most descriptive visual equivalent to the process of fascial hydration through movement.

Try It Out
Envision or grab a household sponge and water.

Scrubby pad: The scrubby pad represents the hydrophobic fibres.

Soft pad: The soft pad of the sponge represents the hydrophilic (water-binding) ground substance.

Mix: In your imagination, intermesh the scrubby fibres within the soft pad. This is a nice image of the fascial fibres enmeshed within ground substance.

Bowl of water: The water represents interstitial fluid.

Squeeze and release to rehydrate fascia

Soak, Squeeze, Rehydrate
Put the sponge into water.

Soak: The sponge soaks up as much water as possible.

Hold: After lifting the sponge out of the water, some fluid remains attached to the material. It represents water bound in fascia.

Squeeze: By squeezing the sponge, water is extruded from the fabric. When the sponge is dipped into the water again while a firm grip is kept on it, very little water is drawn back into the material.

Squeeze to cleanse

Rehydrate: Once you loosen your grip, water is reabsorbed.

Release to nourish

Manual therapies use this principle of squeezing and releasing to 'wash out' pro-inflammatory substances and waste products to enhance tissue nourishment. The technique is also used to dissolve light adhesions of the collagen network to enhance the nutrient and oxygen supply to fascia. In a different way, the same principle can be applied to movement.

HYDRATING MOTIONS

Hydration with movement interplays

In Slings, specific movement interplays are used to promote tissue hydration and fluid flow. By complementing tissue tension or compression with tissue softening, the intention is to squeeze fluid, and with it waste products out and then allow fresh, nutritious fluid to be drawn back into the fascia. Hydrating interplays are mostly used in combination with self-massage and active lengthening exercises; however, the principle can also be applied to toning exercises.

Hydrating Self-Massage

Pressure followed by release

Self-massage exercises with props, such as balls and rollers, are a wonderful way to cleanse and rehydrate fascia. While the pressure of the tool is intended to squeeze fluid out of the tissue, the subsequent release is the rehydration phase.

Every prop has its advantages. In Slings, we mostly use soft, textured massage balls because they allow three-dimensional movement, are suitable for all areas and can reach smaller, less accessible places of the body.

When self-massage exercises are used for tissue hydration rather than stimulation; speed, depth, and body organisation make a difference.

Slow rolling:	Hydrating massage movements are done at a slow pace.
Sustained pressure:	Pressure in one area can be sustained for a short period of time. A duration of ten to twenty seconds is generally sufficient.
Softness for depth:	For the best 'sponge effect' and to reach deeper myofascial structures, the more superficial layers need to be as soft as possible. Ideally, the body is organised in a way that allows the massaged part to be relaxed.
Matching size:	If you want to be specific in your 'touch', match the size of the prop to the body part being massaged.

ITB Massage Zigzag with Massage Ball   Quadriceps Stroke with Massage Ball

Hydrating Movement Interplay: Active Lengthening and Softening

Active lengthening of myofascial structures, which tensions the fascia, is complemented with tissue softening.

The complementing motion in which tissue softens is in the opposite direction of the active lengthening component. For the best effect, the tissue is shortened passively. For example, active hip extension is complemented with 'passive' hip flexion.

Lengthening followed by softening into the opposite movement direction

Kneeling Crescent Lunge & Shift

Hydrating Movement Interplay: Toning and Softening

Muscle strengthening is followed with muscular relaxation and fascial softening.

The complementing position in which tissue eases is in the same direction as the concentric contraction of the prime mover(s). For example, strengthening of the hip abductors is complemented with passive hip abduction and strengthening of the hip adductors is complemented with passive hip adduction.

Toning followed by softening into the same movement direction

Side to Side with Hip Massage

Lower & Lift

Pelvic Tilt

Diamond Legs

Sacrum Opening

PART 2

Potential short-term discomforts worth the long-term benefits

Going with the Flow

As previously mentioned, fascia can become excessively viscous or fibrous when unused or mistreated. The excessive fibrosity or density of the tissue limits the fluid exchange, thereby hindering the removal of waste products and the delivery of nutrients. Introducing movement to dried out, gluey or 'toxic' areas promotes detoxification and nourishment. As positive as that is, there can be some short-term side effects like queasiness, headaches, nausea, or a general sense of malaise or grumpiness. There is no reason for concern; they are just signs that the body is responsive, and that internal cleansing is in progress. Take it easy and rest if need be. Trust that the short-term discomfort is worth the long-term benefits.

HEALTH-PROMOTING TEACHING METHODOLOGY

In addition to using self-massage and hydrating motions to nourish the body's 'vitalising inner ocean', we emphasise the health benefits of fluid flow throughout the fascial system in our teaching methodology.

Communication: With constructive language during our teaching, we encourage trust in the health and healing-promoting qualities of fascia.

Class structure: We believe that inner and outer flows enhance each other. Therefore lessons are functionally choreographed, linking one motion to the next in a seamless manner.

Resource-oriented language and class flow

PART 2

ONTO THE MAT AND INTO THE BODY

▶ Story and Practice: Fluidity in Motion

SLINGS MYOFASCIAL TRAINING TECHNIQUES AND AIMS

The following Slings Myofascial Training Techniques and Training Aims shown in gold, bold font are directly related to fascial fluidity.

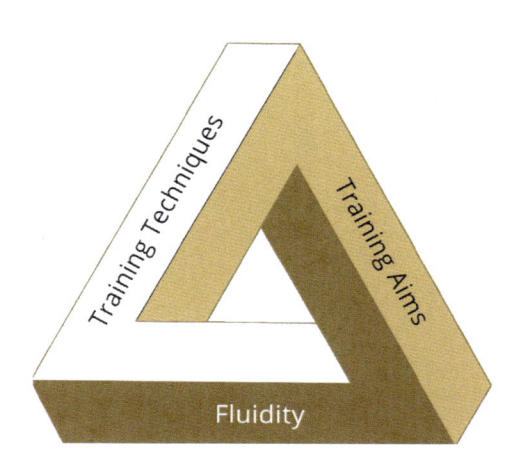

12 Myofascial Training Techniques
Techniques that optimise fascial hydration:

1. Stabilising
2. Toning
3. Pushing, Pulling, Counter-traction
4. Active Lengthening and **Expansion**
5. **Spiralling**
6. **Hydrating**
7. **Gliding**
8. **Domino Motion**
9. **Bouncing and Swinging**
10. **Massaging**
11. Active Ease
12. Melting and Invigorating

12 Training Aims
Four top achievements of fluid fascia:

1. Dynamic Stability
2. Multidimensional Strength
3. Limberness
4. Movement Rhythm
5. Elasticity
6. **Tissue Nourishment**
7. **Adaptability**
8. **Resilience**
9. Somatic Trust
10. Movement Courage
11. **Mañana Competence**
12. Kinaesthetic Intelligence

IN A NUTSHELL
THINGS WORTH KNOWING ABOUT FASCIAL FLUIDITY AND MOVEMENT

3 Things About Extracellular Matrix

Extracellular matrix:	Fascia is fluid-filled extracellular matrix (ECM), the watery, living environment of cells.
Cellular health:	The health of cells greatly depends on the health of the immediate living environment.
Fascial health:	Sufficient hydration and fluid flow support fascial health, and as a result, vitality.

3 Things About Fascial Hydration

Versatile movement:	In general, versatile movement promotes fluid flow and fascial hydration.
Limiting factors:	Factors contributing to restricted fluid flow in fascia are gluey (excessively dense) ground substance, felty (excessively fibrous and disorganised) patches, and adhesions.
Sponge effect:	Fascial hydration is actively supported by utilising the 'sponge effect': compressing and releasing or tensioning and softening the tissue.

3 Slings Training Principles

Sponge principle:	We use the 'sponge effect' in hydrating interplays without props and in self-massage exercises to promote fluid flow in fascia.
Pressure, release:	Nourishing self-massage exercises are done slowly with the treated body part as relaxed as possible. Pressure can be sustained for 10 – 20 seconds in one area.
Lengthening, softening:	In active lengthening exercises, the lengthening is followed by softening into the opposite movement direction.
Toning, softening:	In toning exercises, the contraction is followed by softening into the same movement direction.

3 BENEFITS OF FASCIAL FLUDITIY

1. Vitality
2. Self-healing
3. Movement ease

PART 2

The **glide** in fascia facilitates your movement ease.

7. GLIDE

Fascia is a gliding system in which the loose, watery tissue provides sliding layers between other myofascial structures. Glide facilitates movement ease and body awareness.

Let Me Ask You

What makes your body feel lithe and your movements fluid, versus unyielding and laboured? The same thing that makes you perceive motion immediately and accurately: fascial glide.

The well-lubricated loose fascia forms a bodywide sliding system. It provides gliding layers that enable neighbouring myofascial structures to slide against each other in a dynamically stabilised, yet fluid manner.

A Good Reason To Care

If you prefer cat-like suppleness over movement labour, you have good reason to get onto the mat regularly to care for your fascial glide.

PART 2

BODYWIDE SLIDING SYSTEM

Loose fascia is where glide occurs. Because glide is of the utmost importance where there is motion, loose fascia is omnipresent in the body. Before looking into fascial glide and related movement strategies, let's revisit the fasciacytes to fortify the focus on loose fascia and remember the importance of fascial hydration.

Fasciacytes

Devoted to hyaluronan production

The recently discovered fasciacytes are cells devoted to the production of hyaluronan, an important glycosaminoglycan (GAG) of the ground substance that binds water and supports fascial hydration. Ground substance itself forms the lubricating gel that facilitates smooth sliding motions within the fascial system.

Lubricating gel

Fasciacytes are especially present in the interfaces between fascial layers where glide is functionally required. For example, between superficial and deep fascia, neighbouring deep fascial structures, deep fascia and the epimysium of muscles, neighbouring epimysia, and the sliding fascia within a muscle. For hyaluronan and other GAGs to be able to absorb water, there needs to be sufficient fluid in the body. This fluid also needs to get into the tissue, not just flush through your digestive system, as discussed in the chapter about 'Fascial Hydration'. Movement in general, and generous, versatile movement, in particular, are supportive of fasciacyte stimulation, therefore optimal fascial hydration and glide.

Martina, Alexa, and Karin lubricating in the Indian Ocean on South Beach in Western Australia

LOOSE FASCIA

Compared to deep fascia, loose fascia has a high fluid and low fibre ratio. Its fluidity and multidirectional fibre organisation make it an ideal gliding tissue for neighbouring myofascial structures.

High fluid, low fibre ratio

Viscosity

Viscosity is a measure of fluid resistance. In common parlance, a liquid is viscous when it is denser than water. Examples of viscous substances are syrup, body lotion, and the kind of mucus secreted by snails, the latter being fairly close to the consistency of ground substance (although you might want to visualise differently).

Viscous substances are denser than water

Variable Fluidity

Remember that ground substance is a gel that can be more fluid or dense. The degree of its viscosity naturally varies in different parts of the body, based on functional requirements.

Degree of viscosity naturally varies

Low viscosity: Ground substance with lower viscosity is more fluid, and therefore facilitates sliding motions more readily.

High viscosity: Ground substance with higher viscosity is denser, and therefore more resistant to change induced by movement.

Functional Aspects of Viscosity

Three main movement-relevant aspects of fascial viscosity are glide, movement deceleration, and shock absorption.

Glide: The viscosity of the ground substance facilitates frictionless sliding motions between collagen fibres and smooth glide between adjacent myofascial structures. Besides increasing movement efficiency by decreasing resistance, fascial glide also enhances kinaesthesia, which, in turn, enhances movement coordination.

Deceleration: Together with fibres, the gel-like nature of the ground substance slows down length changes in fascia. For example, in movements where muscles work eccentrically (lengthen in an active manner), the decelerating quality of fascia conserves muscular energy and therefore increases movement ergonomics.

Shock absorption: The viscosity of the ground substance also functions as a shock absorber. As a fluid buffer, it protects and supports the longevity of joints and other structures.

STABILISED GLIDE

Relative motion between adjacent surfaces

Essentially, glide means that two adjacent myofascial structures slide relative to one another, using the loose fascia as their gliding medium.

While the ground substance is the lubricant, it is the fibres that control the degree of glide. The multidirectionally-organised collagen and elastin fibres flexibly link adjacent deep fascial layers, which provides dynamic stability as the surfaces slide against each other. In this way, the range of gliding motion is contained, and structural integrity is maintained.

Contained range of gliding

Variability of Glide

Great variability in range, direction, and speed of gliding motions

There is great variability in terms of the range, direction, and speed with which individual layers travel when we move. Because the principle of 'form follows function' applies, not all myofascial structures have or need the same degree of sliding. Functionally, there is or should be a lot of glide between a retinaculum and the tendons beneath to allow full range of motion in the wrists, elbows, knees, and ankles. In contrast, in the palm of the hand and the sole of the foot, the skin, superficial fascia, and the deep fascia beneath are strongly connected. Between these structures, stability is functionally favoured over adaptability. The degree of glide between broad fascial sheaths, such as the abdominal aponeuroses or the thoracolumbar fascia, is somewhere in between.

Functional variability in the degree of glide in different body parts

Lubricating gel can become adhesive glue

Gluey Gel
In addition to a healthy fluid intake and a nutrition that supports an alkaline (non-acidic) milieu in the body, movement is key to optimal tissue hydration and, therefore, fascial glide. Imposed or chosen inactivity negatively influences fluidity of the ground substance, and the gel can become glue.

Turning glue into lotion

Adhesive glue can become lubricating gel

Ungluing Fascia
While different degrees of glide are natural, being glued together on the inside is clearly unhelpful. Not only does it make activity more laboured and disintegrated, it also decreases movement love and the rate with which the body rejuvenate. Besides, it just doesn't feel good to be physically stuck. The good news is, more glide and therefore somatic ease is within reach! More about that later in the chapter.

MOVEMENT EASE WITH INTEGRITY

If not intellectually, somatically, it is abundantly clear that fascial glide is a key ingredient for movement ease. Not only does it prevent fascial restrictions, it also lubricates the body from within. Another quality that has not been mentioned yet is enhanced kinaesthesia. This is an important sensory aspect in which gliding motions refine movement coordination along with body awareness.

Glide facilitates movement ease, both mechanical and sensory

Lucas and Karin coordinating relative motion in Berne in Switzerland

Healthy Movement Differentiation

Much of the time and for good reason, we speak about the importance of fascia as a connector. However, fascia is also a differentiator that enables individual muscles to function efficiently, without restricting their neighbours (remember the part about excessive force transmission in chapter 3). Glide between fascial layers can be seen as a form of healthy tissue differentiation. This differentiation is not only essential for optimal movement functionality, it is also crucial for dynamic stabilisation that truly comes from within. It is the kind of inner stability in which local muscles work at different times to different degrees without being suppressed by excessive fascial tension created in overactive outer muscles.

Myofascial differentiation for optimal dynamic stability and movement

Kinaesthetic Intelligence

Movement coordination and body awareness

Loose fascia is rich in sensory receptors that are stimulated by sliding motions. This means that fascial glide strengthens kinaesthetic intelligence, therefore proprioceptive finesse and interoceptive clarity. In other words, refined movement coordination and clear perception of how we feel (more about that in chapter 11. Kinaesthesia).

Fascial Restrictions in Disguise

'Stuck' fascia can be confused with muscle dysfunctions

When functional fascial glide is restricted, it can have a series of negative consequences that are often challenging to distinguish from muscle dysfunctions. What might be perceived as muscle tension could actually be a fascial restriction that inhibits optimal muscle functionality.

Movement Restriction

Muscle lengthening might be hindered

Depending on the degree of excessive connectedness between myofascial layers, movement is either more effortful, somewhat restricted, or almost completely inhibited. At times, this happens to all of us somewhere in the body. And now contemplate how often you hear something like this, "I think I have a fascial restriction in my hamstrings; therefore I would like to improve glide between and within them", as compared to, "my hamstrings are tight, I really need to stretch them". We have learned to associate certain feelings with certain causes, and, in the realm of movement, muscles are believed to be a likely cause for all sorts of issues. What if they are not, or not exclusively? Then, the hamstring stretch doesn't have the desired or long-lasting effect. A fascia-focussed Plan B is needed!

Excessive Force Transmission

Muscle relaxation might be inhibited

Adhesions between myofascial structures that should have a degree of differentiation or glide, can lead to excessive force transmission between them. Meaning, the activity of one muscle can tension the epimysium (outer muscle fascia) of its neighbour and therefore increase its resistance to change. With this in mind, maybe the over-recruitment patterns we frequently see in core stabilisation training or the inability to relax synergistically working muscles are not a neuromuscular coordination issue, but a fascial adhesion issue. If this is true, at least in part, different strategies than the traditional neuromuscular-focussed exercises need to be applied to improve overall functionality sustainably.

Lowered Kinaesthesia

When functional glide is restricted, kinaesthesia and with it, neuromuscular movement coordination and body awareness are lowered, at least locally. While stretching exercises have their benefits, they don't stimulate receptors in the same way as sliding motions. Hence, there might be sufficient flexibility in muscles, yet areas of restricted fascial glide and limited kinaesthesia. To utilise the movement ease that is potentially available in the body, gradually becoming unstuck between muscles is essential.

Neuromuscular coordination and body awareness might be impeded

Fascia-Focussed Movement Remedies

Regardless how stuck the body feels, something can be done to have more ease. In addition to the glide-enhancing practical applications shown later in this chapter, the movement strategies discussed in previous and upcoming chapters are supportive of turning fuzzy fascial glue into a more fluid gel. The chapters are:

- 1. Tensile Strength
- 3. Force Transmission
- 4. Adaptability
- 5. Multidimensionality
- 6. Fluidity
- 9. Plasticity
- 10. Tone Regulation
- 11. Kinaesthesia

The movement strategies in the chapters just mentioned, in combination with the remaining ones, facilitate the maintenance of functional glide. The remaining chapters are:

- 2. Muscle Collaboration
- 8. Elasticity

All Fascial Movement Qualities are utilised to unglue fascia

PART 2

Martina and Karin gliding on a slide in Fremantle in Western Australia

POSITIVE-SELF-SUSTAINING CYCLE

Fascial glide supports multidimensional movement ease by enhancing tissue hydration and healthy differentiation, movement coordination, and body awareness. In turn, multidimensional movement ease sustains fascial glide. A positive, self-sustaining cycle is set and kept in motion.

Fascial glide and movement diversity create a lubricating cycle

Fluid Ground Substance

Fluidity: The fluidity of loose fascia facilitates glide.

Motion: Gliding motions between myofascial structures stimulate cells that lubricate loose fascia.

Dynamically Stabilising Fibres

Fibre organisation: The loosely knit, multidirectional fibre organisation in loose fascia permits glide.

Motion: Gliding motions sustain the loosely knit, multidirectional fibre organisation in loose fascia.

Fibre looseness: The fibres in loose fascia dynamically stabilise adjacent surfaces, thus facilitating structural integrity.

Motion: Gliding motions 'melt' excessive fibres in loose fascia, thereby supporting dynamic stability and preventing structural fixation.

Multidimensional Movements

Multidimensionality: The different speeds and directions in which myofascial layers are able to glide support movement diversity.

Motion: Versatile movement engages loose fascia in diverse ways, with its functional multi-dimensionality.

Kinaesthesia

Kinaesthesia: Sliding motions stimulate sensory receptors that enhance movement coordination and body awareness.

Motion: Well-coordinated movement and body awareness facilitate activities that utilise and sustain glide.

SLINGS IN MOTION

When we move, myofascial glide happens throughout the body; within muscles, between muscles, and between fascial layers. In terms of practically applying this knowledge, we unite two approaches.

- Improving glide more or less indirectly by utilising all of the other Fascial Movement Qualities.

- Enhancing glide deliberately as described in this chapter.

To facilitate glide directly, we differentiate exercises that aim to enhance glide from exercises that utilise existing glide. As conceptual and subjective this differentiation may be, it has proven cognitively useful and somatically successful in practice, therefore it is a valuable distinction to consider.

EXERCISES ENHANCING GLIDE

Improve and sustain glide

With glide-enhancing movements and massages, we aim to sustain existing glide or gradually improve it in myofascial structures that are glued together or lightly adhered.

Glide-Enhancing Movements

Glide between deep fascial structures and muscles

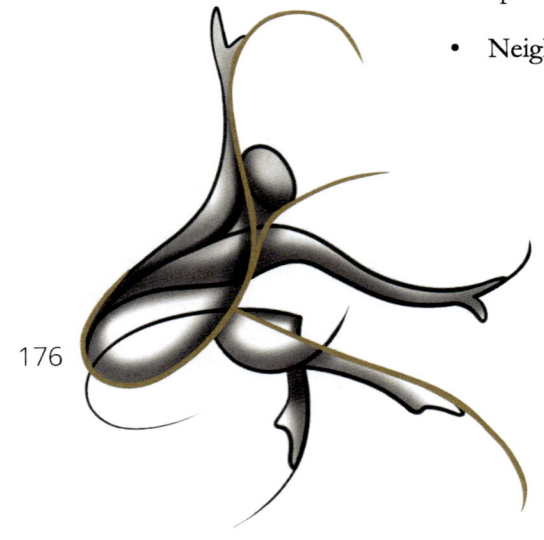

These kinds of movements have such unique effects that it is hard not to put an exclamation mark in the header. Although they work within muscles as well, the exercise focus is on relative movement between:

- **Deep fascial structures** (such as tendons, ligaments, and **apo**neuroses)

- **Nei**ghbouring muscles (their epimysia)

Common Features

Common features of glide-enhancing exercises are a generous range of movement (ROM), appreciable tension changes in layers, and moderate intensity.

Generous movements with medium intensity

Generous ROM: Ideal is a range of movement that is large enough to create perceptible glide between myofascial layers.

Tension changes: Ideal are unidirectional or multidirectional movements that create noticeable tension changes between adjacent myofascial layers.

Moderate intensity: Ideal are movements that allow variations in loading. A perceived forty to sixty percent of maximum load (muscle contraction and fascial tension) works well in practice.

Sit Back & Curl Up into Arch

Spirals, Arcs, Curling Waves, and Dominos

Spirals, arcs, and wave-like movements, as well as 'domino' motions, are excellent to promote glide. Hence, they are frequently part of glide-enhancing exercises.

Multidimensional movements

Spirals: Actual or perceived spiralling motions in the spine, hip, and shoulder joints create glide in and around the moving joints.

Arcs: Generous arc-like movements of the trunk, arms, and legs promote glide between broad fascial sheaths around the spine and across the hip and shoulder joints.

Curling waves: Curling and wave-like motions of the spine, including pelvic movement, enhance glide between the myofascial structures in the upper body.

Domino: Domino motions involving the whole or part of the body facilitate glide from the outside-in, from superficial to deep fascia to muscle fascia.

Adductor Stretch & Glide with Wave

PART 2

Glide between superficial and deep fascia

Slow, relaxed rolling

Low Kneeling Calf Massage & Pelvic Shift with Massage Balls

Glide-Enhancing Massage
For superficial glide, slow self-massage exercises work beautifully. The exercises focus on relative movement between:

- Superficial fascia and deep fascia

Common Features
During the massage, the skin and superficial fascia are rolled over deeper myofascial structures. This enhances glide between the superficial fascia and deep fascia, and increases fluid flow and sensation.

Slow pace: A slow pace is ideal for glide-enhancing massages.

Low repetitions: Generally, a very low number of repetitions is sufficient; between one and three strokes.

Softened tissue: Ideally, the body part that is massaged is as relaxed as possible to decrease tissue tension.

Require glide for efficient, pain-free execution

High intensity, relatively small range of movement

Thigh Stretch

EXERCISES UTILISING EXISTING GLIDE

Based on experience, some exercises don't feature the degree of fascial adaptability, fluidity, and range of movement that is necessary to enhance decreased glide. Instead, they require pre-existing glide to be performed effectively and pain-free.

Common Features
Exercises that utilise pre-existing glide often feature high loading on muscles and fascia, in combination with relatively small tensional differences in adjacent fascial structures.

Strong contractions: The prime movers are strongly activated to perform the exercises. The perceived intensity is eighty percent and upward.

High tension: The involved fascia structures are highly tensioned, thus have comparatively low adaptability.

Small variations: Tension changes in adjacent fascial layers are relatively small.

SLINGS MYOFASCIAL TRAINING TECHNIQUES AND AIMS

The following Slings Myofascial Training Techniques and Training Aims shown in gold, bold font are directly related to fascial glide.

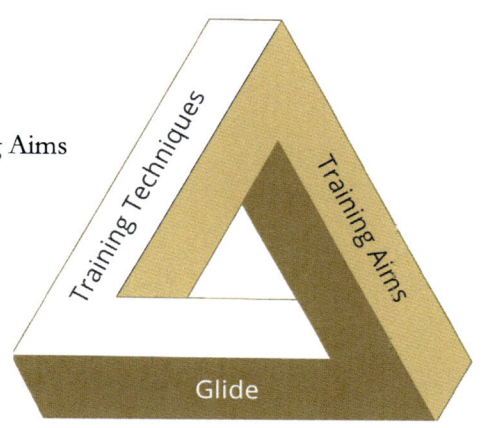

12 Myofascial Training Techniques
Techniques that optimise fascial glide:

1. Stabilising
2. Toning
3. Pushing, Pulling, Counter-traction
4. Active Lengthening and **Expansion**
5. **Spiralling**
6. Hydrating
7. **Gliding**
8. **Domino Motion**
9. Bouncing and Swinging
10. Massaging
11. Active Ease
12. Melting and Invigorating

12 Training Aims
Four top achievements of fascial glide:

1. Dynamic Stability
2. Multidimensional Strength
3. **Limberness**
4. Movement Rhythm
5. Elasticity
6. **Tissue Nourishment**
7. **Adaptability**
8. Resilience
9. Somatic Trust
10. Movement Courage
11. Mañana Competence
12. **Kinaesthetic Intelligence**

ONTO THE MAT AND INTO THE BODY

Story and Practice: Glide in Motion

IN A NUTSHELL
THINGS WORTH KNOWING ABOUT FASCIAL GLIDE AND MOVEMENT

4 Things About Loose Fascia

Loose fascia:	Loose fascia is omnipresent in the body; wherever there is motion, there is loose fascia.
Favouring fluidity:	The high fluid and low fibre ratio in loose fascia makes it an ideal sliding medium.
Viscosity:	The viscous ground substance facilitates frictionless glide within and between muscles and fascial structures.
Dynamic stability:	The loose, multidirectional fibres that connect adjacent surfaces permit a degree of glide while dynamically stabilising tissue layers.

4 Things About Glide

Functional variability:	There are functional variations in the range, direction, and speed of gliding motions in different body regions.
Movement ease:	Glide facilitates movement ease by way of preventing fascial restrictions, lubricating the tissue, and enhancing coordination along with body awareness.
Glide restrictions:	When fascial glide is restricted, it can have a series of negative consequences that are often challenging to distinguish from muscle dysfunctions.
Glide enabling:	Generally, by utilising all of the Fascial Movement Qualities and specifically, by practising glide-promoting exercises, glide can be gradually retrained and sustained.

4 Slings Training Principles

Glide-enhancing moves:	Common features of glide-enhancing exercises are a generous range of movement, appreciable tension changes in adjacent layers, and moderate intensity.
Spirals to dominos:	Spirals, arcs, curling, wave-like movements, and domino motions are features frequently incorporated into glide-enhancing exercises.
Glide-enhancing massage:	Common features of glide-enhancing self-massages are rolling in a slow pace, with low repetitions and the treated body part as relaxed as possible.
Glide-utilising moves:	Common features of glide-utilising exercises are high loading on muscles and fascia, in combination with relatively small tensional differences in adjacent fascial structures.

3 BENEFITS OF FASCIAL GLIDE

1. Movement ease
2. Fascial hydration
3. Movement coordination and body awareness

PART 2

The **elasticity** of fascia adds buoyancy and efficiency to everyday movements and athletic activities.

8. ELASTICITY

Fascia is an elastic tissue that adds spring and buoyancy to oscillating motions. Elasticity contributes to efficiency, power, and movement love, which is invaluable for body and spirit.

Let Me Ask You

How does a child tirelessly skip rope? What gives a downhill skier inner spring suspension? From where do some bodies get their walking lightness? Or in contrast, why does shuffling the feet seem to be an inevitable outcome of aging (which it is not in fact)? Fascial elasticity, or the lack of such, contribute to answering all of these questions. With its uplifting quality, elasticity is key to movement love, efficiency, and ease at any age.

A Good Reason To Care

Every step you take is a good reason to get onto the mat and take care of your elastic movement capacity. Fascially empowered athletic performance is just an added benefit.

THE BODY'S ELASTIC MOVEMENT CAPACITY

Fascia stores and releases movement energy elastically

Fascia has the capacity to store and release kinetic (movement) energy. Meaning, it can elastically lengthen and recoil to add spring and buoyancy to everyday movements as well as athletic activities. Most amazingly, while it gives you greater movement ease and power, it also saves you energy. Fascial elasticity is a physically and emotionally uplifting quality that is worth training or retraining!

A Word About Wording

To avoid linguistic confusion, let's speak about the terminology used in this chapter before jumping into the topic.

Tissue Elasticity and Resilience

You will come across the term 'resilience' in this and the chapter about the twelve Slings Training Aims, which makes it important to distinguish the different meanings.

In physics, resilience is the ability of an elastic material (such as an exercise band or a tendon) to absorb energy and release that energy as it springs back to its original shape. Regarding fascial elasticity, the terms 'elasticity' and 'resilience' are, therefore, interchangeable.

In a different context, resilience stands for the inner strength to physically and emotionally recover from adversity, even emerge from challenging times stronger than before. This kind of resilience is referred to in the twelve Slings Training Aims.

Elastic Lengthening and Recoiling

When referring to the elastic behaviour of fascia, I will use the following terms, because I find them the most descriptive and easy to grasp:

- Elastic lengthening
- Recoiling

The expression 'elastic lengthening' is interchangeable with other terms you might be familiar with:

- Elastic tensioning
- Elastic loading
- Pre-tensioning

Fascia's Elastic Energy Contribution to Movement

By increasing movement power and reducing the energy expenditure in muscles, the kinetic storage capacity of fascia increases performance in activities like walking, running, leaping, bouncing, and rhythmical swinging motions, as well as energetic throws and kicks. To have an easy to understand image of what kinetic storage capacity encompasses, visualise an elastic band.

Elastic lengthening: When elastically lengthened (tensioned), fascia stores kinetic energy, similar to an elastic band that is stretched.

Recoiling: When the tension is released immediately after the elastic lengthening, the kinetic energy is freed, and the tissue recoils naturally, similar to a stretched elastic band that snaps back.

Kinetic storage capacity encompasses elastic lengthening and recoiling

FUNCTIONAL ELASTICITY

To spring or not to spring, that is the functional question.
To embrace the wonder of elastic movement, it is worth briefly looking at functionally non-elastic movement.

Non-Elastic for Good Reason

Some movements are functionally non-elastic

For good reason, not all of our movements are elastic. Purpose-driven, non-oscillating activities, such as typing the words on this page or sipping a tea, are non-elastic and coordinated from the 'top-down': brain to muscles, fascia, and bones.

Steady Movement Coordination

Interoceptive sensation and/or intention
↓
Nervous system
↓
Muscles
↓
Fascia
↓
Bones

For example, when you are thirsty (interoceptive sensation), and you want a drink (intention), you reach for a glass of water (the central nervous system relays the information to the somatic part of the peripheral nervous system, which activates the necessary muscles, whose contractions tension fascia, which subsequently moves bones), and finally the glass arrives at your mouth.

GUCCI
La Maison
ne fait
Pas de Crédit
GUCCI
PÉRITIF A L'ORA
CARMINI

PART 2

Karin proactively training and retaining her fascial elasticity on the Avenue of Stars in Hong Kong

Oscillating movements are functionally elastic

Elastic for Good Reason

Fascial elasticity mostly comes into play when gravity and movement momentum interact; add ground reaction force, and potentially, you have a very energy-efficient, springy synergy.

Dynamic Movement Organisation
Following is a simple model for envisioning the sequence of events in elastically empowered movements, such as walking and running.

Gravity and movement momentum → Ground reaction force → Bones → Fascia and muscles → Nervous system → Muscles and fascia → Bones

Let's say you are walking in your preferred rhythm. Because of gravity, you 'fall' with every step. The joints in your lower body fold in a spiralling fashion. The associated fascia starts decelerating the movement. The changes in fascial tension and muscle length stimulate mechanoreceptors. They inform the central nervous system (spinal cord) of what's going on, which triggers a muscle response (a relatively isometric contraction). At the same time, the fascia stores energy by elastically lengthening. When sufficiently tensioned, the stored energy is released and fascia recoils, which moves joints in the opposite direction, thereby propelling you forward …

Note that the first or first few steps you take are different. They are, as described in the previous section, steady actions in which the nervous system coordinates the respective muscles to set you into motion. To utilise fascial elasticity effectively, continued rhythmical movement is required.

Proactively Reclaiming the Price of Modern Life

To make the energetic difference between continued, elastically-empowered, and discontinued, primarily neuromuscularly coordinated actions more tangible, engage in a mental exercise. Compare the feel of an extensive, uninterrupted hike in nature, with the quality of an urban walk, where traffic lights necessitate stopping and starting frequently. Besides the rejuvenating effects of fresh air and outdoor activity, it might also be the energy-efficient utilisation of your fascial elasticity that invigorates your being and enables you to be active for hours without feeling drained. Versus the weariness many of us experience after just one hour of crossing streets and browsing shops.

In contrast to our ancestors, who sustained their resilience with extensive walks on natural grounds, most of us primarily engage in the more energy consuming stop-start actions of modern life. Often, the resulting gradual loss of movement spring is then blamed on ageing – end of story. At this point, it is beyond important to remember that you are the author of your fascial story, and that regained and sustained buoyancy can be part of it!

PART 2

Describes fascial and muscular interplay in elastic movements

Catapult Mechanism

To appreciate the current scientific understanding of fascial elasticity, it is worth taking a glance at its origins. The ability of fascia to store and release kinetic energy was first documented when scientists investigated how kangaroos perform their incredible jumps. They discovered that the contractions of the kangaroos' leg muscles are nearly isometric, while the fascial elements (predominantly the tendons) elastically lengthen and recoil, acting like a spring. They coined this phenomenon the 'catapult mechanism'. Not surprisingly, it was determined other animals, like lean gazelles and tiny frogs, exhibited the same behaviour of utilising their fascia's elasticity for their otherwise unexplainable leaping power. Surprisingly or not, it is now scientifically validated that our fascia has this same kind of capacity; we simply need to use our human version.

Relatively isometric muscle contraction and fascial elasticity

Catapult Mechanism at Work

In the catapult mechanism, the involved muscles are relatively isometrically contracted, meaning the muscles stiffen temporarily without significant change in length. In contrast, the fascial elements elastically lengthen (tension) and recoil (shorten) like an elastic spring, propelling the body forward.

Elastic movements are myofascial, yet fascia is favoured

Muscle contribution: It is important to note that fascia empowered elastic movements still require well-toned, responsive muscles.

Favouring fascia: When it comes to the energy efficiency of these movements, it is the ratio between muscular activity and fascial elasticity that makes the difference. When fascial tissue is sufficiently elastic and tensioned in a suitable rhythm, the muscle-fascia ratio favours fascia. Meaning the movement power is greater without muscles working harder.

Image by Pen Ash from Pixabay.com

RESILIENT FASCIAL ARCHITECTURE

Fibre organisation determines elasticity

Resilient collagen architecture is key

While the components of fascia provide the necessary ingredients, it is the collagen architecture (fibre organisation) that is essential to optimal elastic functioning. At this point, it is good to remember that fascial architecture can be remodelled and that you are the architect of your body.

Viscoelasticity

Quality of ground substance contributes to elasticity

To make the most of the kinetic storage capacity of fascia, the training focus is on (re)building and sustaining a resilient collagen architecture. Understand, though, that both the fibrous (collagen and elastin) and the viscous (ground substance) components of fascia exhibit elastic behaviour. It is their interplay that facilitates best tissue resilience. For uninhibited tensioning and releasing of the fibres during oscillating movements, the ground substance must have the right degree of viscosity. In practice, we attend to the fluidity of the ground substance with hydrating and glide-enhancing exercises while training fascial elasticity directly with more fibre-focussed, springy activities. What is also important to understand about viscoelasticity, is that speed affects tissue stiffening. The faster the loading, the stiffer the tissue becomes. Movement speed or rhythm is therefore an important factor in training and utilising fascial elasticity. If the speed is too slow or too fast, the tissue tension is low or too high respectively for optimal elastic loading.

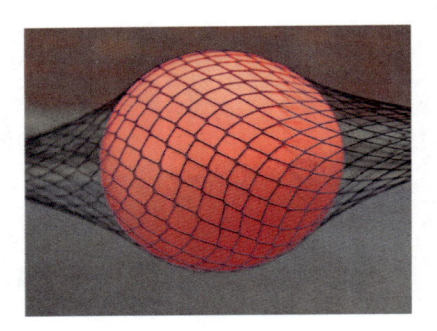

Elastic Collagen Architecture

Lattice
Collagen architecture responds to the way it is loaded. Muscle fascia that is regularly elastically engaged features a two-directional lattice arrangement in its collagen fibre network. Collagen of 'elastically neglected' fascia is much more disorganised (felty) in its orientation.

Crimp
Elastic fascia at rest (not tensioned beyond its natural state) shows a wavy collagen alignment that is called crimp. When fascia is lengthened, the collagen fibres straighten in the direction of the pull until fully tensioned. In this phase, the tissue stores kinetic energy. When the force is released, it recoils to its original length, and the collagen fibres resume their undulating alignment. In this phase the stored energy is released. Crimp formation is essential for utilising fascia as an elastic spring in rhythmical, oscillating movements.

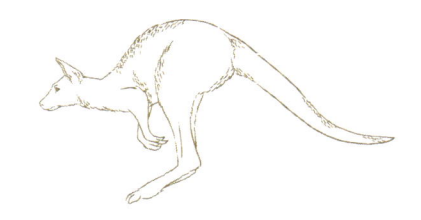

Sustaining or Retraining Fascial Elasticity

Form follows function; function alters with form.

The elastic architecture of fascia can be retrained, optimised, and sustained with movements that feature – yes, you are right – bouncing, jumping, swinging, and elastic spiralling.

What makes and keeps fascia elastic is engaging it in elastic activities – at any age with care, consideration, and a healthy dose of somatic trust and movement courage.

An elastic fascial architecture can be (re)trained

Tom and Karin having elasticising fun in Berne in Switzerland

PART 2

SLINGS IN MOTION

ELASTICITY FOR EVERYDAY FUNCTIONALITY

The Slings repertoire for fascial elasticity primarily aims to support everyday functionality and sportive activities done for a healthy fitness level or for fun. Of course, the exercises can and should be adapted to accommodate specific needs and wants; toned down for differently-abled bodies, and amplified for increased athletic performance.

Where, what, and how is carefully considered

Selection: We are choosy with the myofascial structures we intentionally elastically load, and the intensity and timing with which we train them.

Progressive process over months

Strategy: When training elasticity, we use a progressive strategy. Over the timeframe of several months, exercises advance from simpler to more complex, lower to higher intensity, and from unidirectional to multidimensional.

Upper body: 8 – 15 reps
Arms and legs: 15 – 40 reps

Repetitions: Experience has shown that a low to moderate amount of repetitions is sufficient to gradually enhance and sustain bounce and springiness in the body. Now the question is, what do low and moderate stand for in this context?

Low is between eight and fifteen repetitions in total during a lesson. This range frequently, though not exclusively, applies to upper body movement.
Moderate is about fifteen to forty repetitions in total during a lesson, a range generally suitable for arm and leg movement.

Repetitions taught in sets

No matter the number of repetitions they do not need to be, and often should not be executed all at the same time within a lesson. Instead, consider incorporating breaks between sets, or distribute sets of these kinds of exercises throughout a lesson. Sets may be recurring or different exercises with a similar effect.

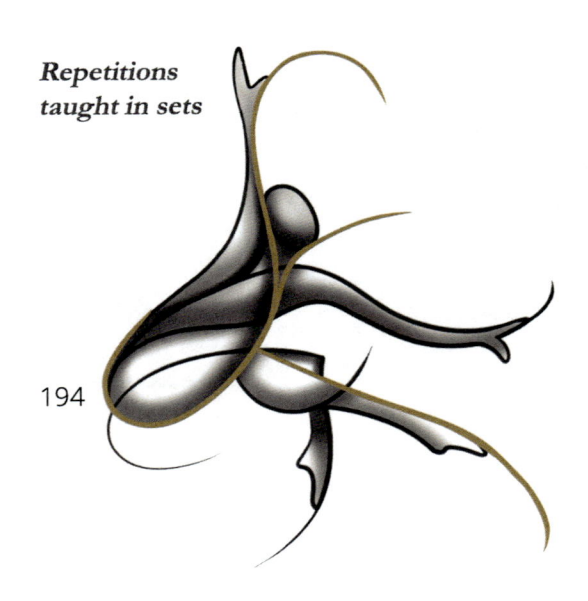

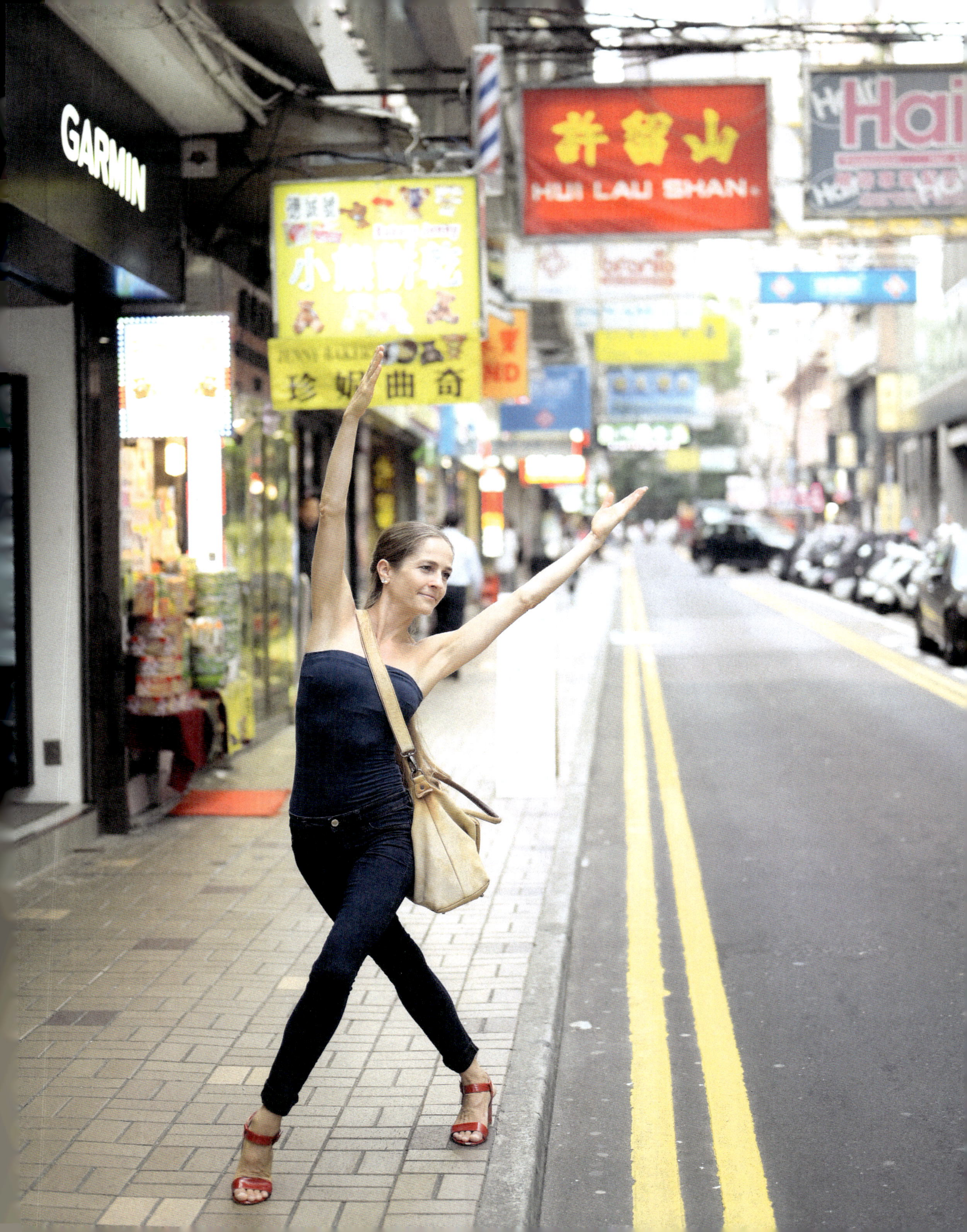
GARMIN
許留山
HUI LAU SHAN

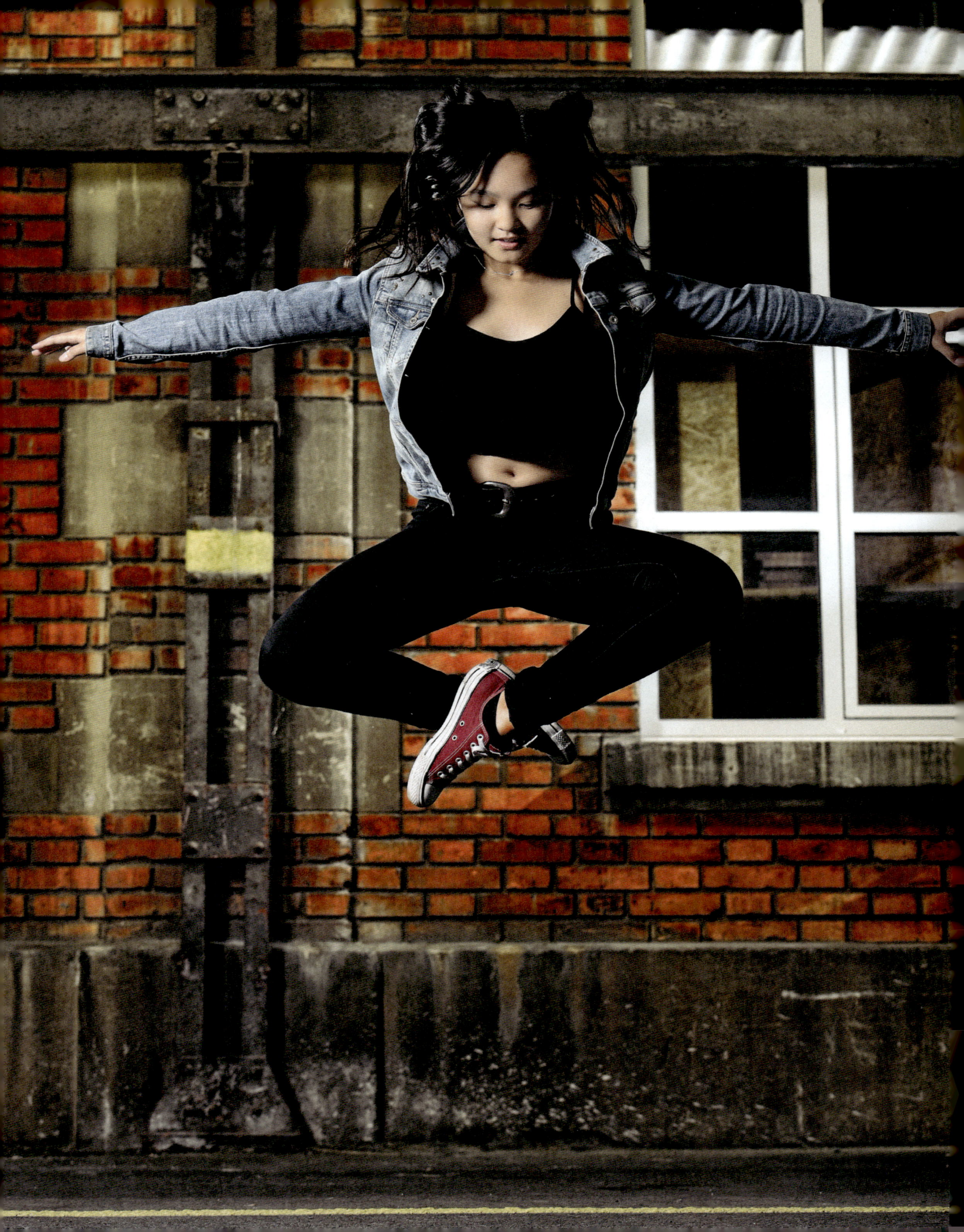

Requirements for Training Elasticity

To utilise and train fascial elasticity, we require one of the following combinations:

- Rhythmical movements against gravity
- Rhythmical movements against gravity including ground reaction force
- Rhythmical movements that elastically lengthen and release fascia with the help of an additional force

Rhythm

Gravity

Ground reaction force

Moving in the Rhythm of (Your) Fascia

The rhythm in which we move significantly influences how efficiently fascia is utilised in oscillating movements.

Oscillating movements in the rhythm of fascia feel springy and buoyant

Motion too slow: When movement is too slow, it becomes more muscularly driven. The muscle-fascia ratio favours the muscles as an energy source.

Transition too long: When the transition phase between elastic lengthening and recoiling is too long, the muscle contribution increases. Once again, in the muscle-fascia ratio, muscles are favoured.

Motion too fast: When a movement is too fast, and fascia can't be sufficiently tensioned, the muscle contribution also increases.

Matching rhythm: When the movement rhythm matches the activity and tensile strength of the engaged fascia, the muscle-fascia ratio favours fascia. The activity feels springy and buoyant.

Whether you walk, run, jump, throw a ball, swing your arms, or dynamically shift and spiral in a Slings class, it is important to find a rhythm that matches your body and the activity. When there is a sense of buoyancy and ease, you've got it!

TRAINING ELASTICITY WITH FAR-SIGHTEDNESS

In Slings, we also think multidimensionally when it comes to training strategies. Instead of 'just' training elasticity, which is a more linear way of thinking, we train its prerequisites to assure safety, efficiency, and enjoyment. In practical terms, we use specific exercises and sequences to:

- Create the conditions for optimal fascial elasticity by focussing on prerequisites

- Train fascial elasticity directly by utilising its inherent resilience

Training Prerequisites and Components of Elasticity

Training prerequisites facilitates safety and sustainability

To train elasticity safely and sustainably, we incorporate exercises that improve the prerequisites and components of fascial elasticity. The training focus is on:

- Movement coordination
- Dynamic stability
- Joint mobility
- Fascial hydration
- Fascial glide

Spiralling Mermaid on Massage Ball

Training Elasticity Directly

In Slings, we train fascial elasticity directly with rhythmical:

- Bounces
- Little jumps
- Swinging movements
- Domino motions
- Elastically-enhanced spirals

Rhythmical movements against gravity or an additional force

Bounces

Bounces refer to dynamic extremity movements in which the feet or hands placed on a fixed surface, such as the floor. Most common are springy knee bends. They are frequently performed in double or steady rhythms.

Dynamic Double Knee Bend & Arm Pendulum

Jumps

Jumps of various sizes are often combined with double bounces.

Inverted V & Dynamic Double Knee Bend with Jump

Swinging Movements

Swinging refers to dynamic, rhythmical arm swings or upper body movements.

- The arm swings are executed in the most non-muscular fashion possible.
- Often, they add momentum to the movement of the rest of the body.
- The rhythm is steady or in harmony with the primary upper body movement.
- Although the swinging upper body motions are elastically enhanced, there is still an element of deliberate neuromuscular coordination for safety.
- In terms of rhythm, exercises often have a slower phase and a faster one. The order of those phases depends on the exercise.

Energy Swing

Domino Motions

In domino motions, one body part follows another with minimal muscular effort. When the technique is applied to elastic exercises, often, one arm initiates the movement, and the rest of the body follows.

Dynamic Hip Release & Side Bend

Elastically-Enhanced Spirals

Elastically-enhanced spirals mostly refer to three-dimensional spinal twists. Unlike bouncing, jumping, swinging, and 'domino-ing' that utilise gravity, elastic spiralling of the spine requires an additional force, typically, one of the arms. Envision a bowstring that needs to be pulled and released by one arm to shoot an arrow. We call this kind of technique the 'bow and arrow principle'.

90/90 Spiralling Twist

Collaborating Forces in Elastic Moves

This table gives you an overview of the external and internal forces that work together in elasticity-focussed exercises.

	Outside the Body		Inside the Body		
	Gravity	Ground Reaction Force	Additional Force	Fascial elasticity	Muscle Activity
Bouncing	✔			✔	As much as necessary, as little as possible
Jumping	✔	✔		✔	Deliberately empowering the jump
Swinging	✔			✔	As much as necessary, as little as possible
Domino-ing	✔			✔	As much as necessary, as little as possible
Spiralling			✔	✔	As little as possible

ONTO THE MAT AND INTO THE BODY

▶ Story and Practice: Elasticity in Motion

SLINGS MYOFASCIAL TRAINING TECHNIQUES AND AIMS

The following Slings Myofascial Training Techniques and Training Aims shown in gold, bold font are directly related to fascial elasticity.

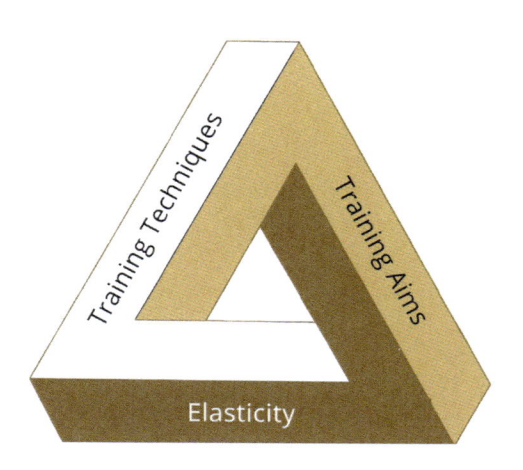

12 Myofascial Training Techniques

Techniques that optimise fascial elasticity:

1. Stabilising
2. Toning
3. Pushing, Pulling, Counter-traction
4. Active Lengthening and Expansion
5. **Spiralling**
6. Hydrating Interplays
7. Gliding
8. **Domino Motion**
9. **Bouncing and Swing**
10. Massaging
11. Active Ease
12. Melting and invigorating

12 Training Aims

Four top achievements of elasticity:

1. Dynamic Stability
2. Multidimensional Strength
3. Limberness
4. **Movement Rhythm**
5. **Elasticity**
6. Tissue Nourishment
7. Adaptability
8. **Resilience**
9. Somatic Trust
10. **Movement Courage**
11. Mañana Competence
12. Kinaesthetic Intelligence

IN A NUTSHELL
THINGS WORTH KNOWING ABOUT FASCIAL ELASTICITY AND MOVEMENT

3 Things About Fascial Elasticity

Kinetic storage capacity:	Fascia has kinetic storage capacity, or the ability to elastically lengthen (tension) and recoil (shorten) to enhance movement efficiency.
Catapult mechanism:	In the catapult mechanism, muscles are relatively isometrically contracted, while the fascia elastically lengthens and recoils to propel the body forward.
Viscoelasticity:	In fascia, both the fibrous (collagen and elastin) and the viscous (ground substance) components exhibit elastic behaviour.

3 Things About A Resilient Fascial Architecture

Resilient architecture:	While the components of fascia provide the necessary ingredients, it is the collagen architecture (fibre organisation) that is key for optimal tissue elasticity.
Lattice and crimp:	Resilient fascia features a two-directional lattice arrangement and crimp formation (wavy pattern).
(Re)Training elasticity:	A resilient fascial architecture can be gradually retrained and sustained with oscillating movements.

3 Things About Elastic Movement

Requirements:	Rhythmical movement against gravity is required to utilise and train elasticity; otherwise, an additional force is needed to elastically lengthen fascia.
Too slow or too fast:	When oscillating movements are performed too slow or too fast, the muscle-fascia ratio favours muscles.
Fascial rhythm:	When the rhythm of oscillating movements matches the activity and tensile strength of fascia, the muscle-fascia ratio favours fascia. The activity feels springy and buoyant.

3 Slings Training Principles

Far-sighted strategy:	Elasticity is trained progressively over months: exercises advance from simpler to more complex, lower to higher intensity, and from unidirectional to multi-dimensional.
Training prerequisites:	To train safely, and sustainably, the prerequisites for fascial elasticity are incorporated into the practice, namely: movement coordination, dynamic stability, joint mobility, fascial hydration, and glide.
Elastic training:	Elasticity is trained directly with bounces, jumps, swinging movements, domino motions, and elastically-enhanced spirals.

3 BENEFITS OF FASCIAL ELASTICITY

1. Buoyancy and spring in every step
2. Elastic movement power
3. Overall physical resilience

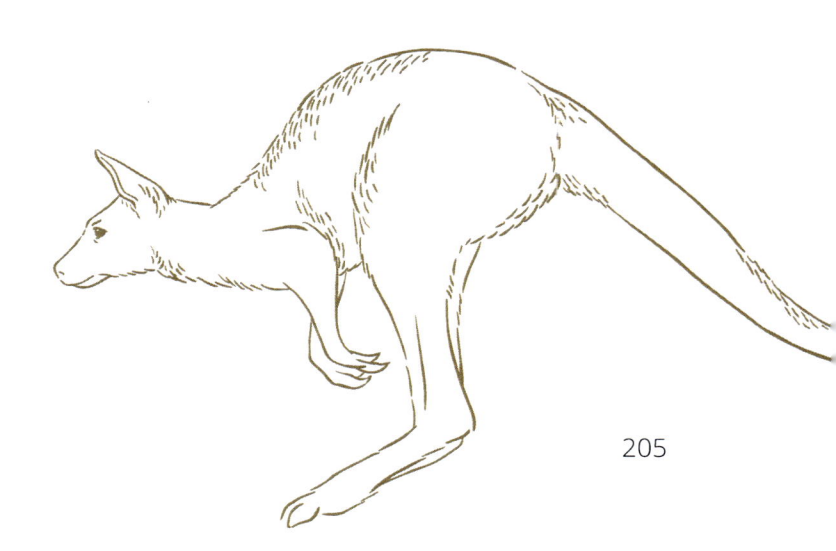

PART 2

The **plasticity** of fascia contributes to your body shape and its lifelong pliability.

9. PLASTICITY

Fascia has natural plasticity, which contributes to the shape of the body and enables long-lasting limberness through practise.

Let Me Ask You
When deeply relaxing into a stretch, where does the delicious sense of melting come from? Fascial plasticity might be the answer you are looking for.

Fascia has natural plasticity, which contributes to, or takes away from your body contours and movement freedom.

A Good Reason To Care
Melted restrictions and freer movements are good reasons to get onto the mat and care for your fascial plasticity in regular, well-measured doses.

PART 2

FORM-GIVING AND SHAPE-CHANGING

"A mind that is stretched by a new experience can never go back to its old dimensions."
Oliver Wendell Holmes, Jr.

Fascia has natural plasticity

Fascia has natural plasticity. In this context, plasticity refers to tissue's ability to give form and, under certain conditions, become pliable to take on a new shape.

Comparing Elasticity and Plasticity
It is helpful to compare elasticity and plasticity to clarify the distinction between these two qualities.

Elasticity: Elastic tissues reversibly deform, like an elastic band that is stretched and released.

Plasticity: Plastic materials irreversibly deform, like a piece of plastic stretched slowly to a new length.

To be clear, this is not to say that with plasticity, deformations are irreversible. The point is that plasticity changes are impactful in a different way than elastic behaviour.

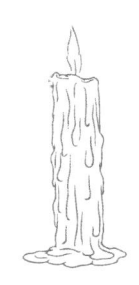

DIFFERENTIATING STATE AND BEHAVIOUR

For a clear understanding of what fascial plasticity is, we distinguish form and function, or state and behaviour.

Form and function, or state and behaviour are differentiated

The term plasticity refers to both the form-giving state of fascia and shape-changing behaviour. In practice, we focus on the shape-changing behaviour, meaning the potential ability of fascia to undergo long-lasting length changes. Referring back to the plastic bag example, we are not interested in the plastic bag itself, but rather the process of changing the shape of the material.

Form-Giving State

Fascia's natural plasticity gives form to individual structures, such as muscles, and contributes to the contours of the body as a whole. Fascia moulds its contents similar to the way the shape of a plastic bottle forms the fluid within.

Gives form to muscles and body contours

Shape-Changing Behaviour

Under certain conditions, plasticity may enable fascia to undergo longer-lasting length changes or alter its consistency temporarily to facilitate immediate alterations with long-lasting effects. Many variables can either permit or inhibit plasticity changes. Influencing factors include, but are not limited to, tissue tone and collagen density, the degree of viscosity of the ground substance, and of course, the emotionally coloured movement behaviour and intentions of the individual mover.

Facilitates longer-lasting shape changes

PART 2

A SENSE OF MELTING

Have you experienced the delicious melting sensation when easing into a relaxed stretch or moving generously and with a sense of pleasure in a pandiculating fashion? What about the sense of yielding in a gently sustained self-massage exercise? If so, have you tried to explain what it is you are experiencing and how or why it works? It isn't so easy, is it? Fascial plasticity has some answers for us. Although the 'why' and 'how' are not yet fully understood, the experience itself is still exquisite.

How Melting May Work

Easeful feedback loop between the nervous, muscular, and fascial systems

The kind of release experienced in melting exercises might be due to the tension or pressure applied to the watery loose fascia. However, it is unlikely to have the same effect on highly collagenous tissues. Most likely, the forces introduced would have to be much greater and held for longer to induce such structural changes. A more likely explanation is that the stimulation of fascial mechano-receptors triggers changes in the nervous system. The enhanced activity in the parasympathetic branch of the autonomic nervous system lowers overall muscle tonus. The muscle relaxation facilitates a quieting of emotional activities and enhances the feeling of inner peace. In turn, muscles soften even more, and so the cycle goes. During this circular, relaxation-inducing interaction between the nervous system and the muscular system, fascia also feels more pliable, possibly due to decreased tension created by muscles and/or a local increase in the fluidity of the ground substance.

Natural Variability and Functional Melt-Resistance

To be clear, in a healthy body, not all fascial structures are meant to undergo plasticity changes with the same readiness. In fact, some will not because they should not alter their length at all. Take, for example, the highly collagenous iliotibial band. No amount of massaging or resting in a melting pose will significantly change this tough fascial structure. Healthy fascia that is functionally sturdy is melt-resistant for good reasons. In the case of the iliotibial band, the reason is the lateral pelvic stability that it provides when standing and walking on two legs.

Structural Reality and Sensory Reality

The sense of melting experienced may or may not be due to actual shape changes in fascia. The good news is, there is equal benefit to structural plasticity changes and a felt sense of melting. Actual fascial lengthening aids tissue rejuvenation locally and can increase the range of movement in associated joints; it is a win. On the other hand, a felt release fosters feeling at ease in the body, which positively influences our movement patterns. Because form follows function, moving in a mañana-competent way remodels fascia in a health-promoting manner; it is also a win. When structural and sensory changes intertwine, we double the benefits: win, win!

Shape-changes may or may not occur

Structural change and the perceived sense of melting are equally valuable

Worn-Out Fascia

There is a significant difference between conscious melting on the mat and habitually wearing out fascia. For example, easing into a sustained stretch that is followed by gentle, active reinvigoration of the muscles and fascia has a profoundly different effect on the fascia of the lower back than habitually slouching in an office chair. The prior is done deliberately, selectively, and within a healthy range. The latter is mostly unintentional and often done within an unhealthy range and timeframe.

Worn-out fascia loses its resilient architecture, leaving the tissue in a disorganised and dehydrated state that decreases its natural glide, elasticity, and kinaesthesia. The constant tension and altered sensation greatly enhance the likelihood of disfunction, discomfort, or pain. This can turn into a vicious 'overstretching cycle' because a person feels they need to stretch the 'tight' area, which is already overstretched. What is needed instead is ease, fluid flow, and gentle glide in the tissues, as well as a gradual rebuilding of a resilient fascial architecture. In short, worn-out fascia doesn't need stretching; it needs softening, rehydration, and invigoration.

Intentional melting and habitual wearing-out are profoundly different

SLINGS IN MOTION

Easing, Yielding, Melting
In this context, the feelings of easing, yielding, and melting are summarised with the term melting.

TIME TO MELT IN HEALTHY DOSES

Melting poses like a delicious dessert: a small dose of the best quality to be slowly savoured.

Benefits include physical and mental ease, limberness, rejuvenation

Melting exercises are a wonderful counterpoise to active, dynamic movements, physically and psycho-emotionally. Besides their contribution to the dynamic balance of a whole lesson, they have an array of immediate and subsequent benefits in their own right. Amongst them are:

- Slowing down
- Consciously letting go
- Physical release
- Emotional ease
- Increased range of movement
- Tissue rejuvenation

Dosage Matters

To avoid adverse effects, less is more

As with most things, it is the dosage that determines if something is nourishment or poison. This is especially relevant for melting poses and massages. For some, not all, if the quantity is too high or the duration too long, these exercises can have a downside. Adverse effects include:

- Feeling of sluggishness and irritation
- Loss of tissue resilience
- Joint destabilisation

If these or other side effects are experienced, don't give up on melting exercises yet! Instead, decrease the exercise quantity and duration, and increase the invigorating motions. Additionally, try adding some muscle-focussed exercises for dynamic stabilisation.

MELTING AND INVIGORATION THROUGH MOVEMENT

In practice, melting and the sense of melting are promoted with:
- Sustained, quiet poses that feel delicious
- Sustained or slow self-massages that feel enjoyable
- Generous, glide-enhancing motions that feel pleasurable

Melting Poses

Poses that gradually lengthen muscles and fascia in a restful manner lend themselves to fascial melting. Melting is not only a fascial affair though. Remember, it is likely a calming interplay of muscles, fascia, and the nervous system. If and in what time frame muscles relax and fascia responds depends on the:

- Body part lengthened
- Overall physical constitution
- Ability to consciously relax
- Response of the autonomic nervous system

Restful poses sustained for 1 to 5 minutes

The average time frame of melting poses ranges from one to five minutes. It is a duration that generally serves the purpose. Therefore it is rarely exceeded. Still, there is no harm in holding some poses for longer, if that is what the body needs.

Melting Deer Pose on Massage Ball

Gentle, dynamic movements

Invigorating Motions

When engaging in melting poses, occasionally even the most mindful movers temporarily destabilise a part of their body or decrease tissue resilience, and therefore physical responsiveness. Both can be prevented without taking away the attained benefits, by complementing melting poses with gentle invigorating motions.

Invigorating motions gently engage and soften the previously lengthened muscles and fascia in an easeful, dynamic manner. The aim is to facilitate a healthy degree of muscular activity and fascial elasticity to retain, or if need be, regain optimal myofascial responsiveness.

Transition:
Melting Deer Pose into Z-Sit

Mermaid Side to Side

Curl in Z-Sit
Transition: Z-Sit into Tailor's Sit

Easeful slow or sustained massages

Melting Massages

In melting massage exercises, easeful slow-rolling, or a sustained comfortable degree of pressure can also create a wonderful sense of melting. Soft massage balls or a roller are often more conducive for the purpose of yielding than hard or edgy props. The duration depends on the:

- Body part massaged
- Density and texture of the massage prop
- Kind of movement or pose

Time frames vary considerably, ranging from ten to twenty seconds, and up to a few minutes.

Z-Sit on Massage Ball

Invigorating Motions

Generally, sustained melting massages soften the area where the massage is applied, while lengthening myofascial tissue elsewhere. For these reasons, subsequent invigoration is recommended.

Curl in Tailor's Sit

Melting Motions

Slow, generous movements that feel pleasurable are another wonderful way to gradually melt inner restrictions. These glide-enhancing motions often comprise multidimensional spiralling, circling, and wave-like elements.

Pleasurable, generous movements that enhance glide

Sitting Hip Opener on Massage Ball

ONTO THE MAT AND INTO THE BODY

▶ Story and Practice: Plasticity in Motion

SLINGS MYOFASCIAL TRAINING TECHNIQUES AND AIMS

The following Slings Myofascial Training Techniques and Training Aims shown in gold, bold font are directly related to fascial plasticity and melting.

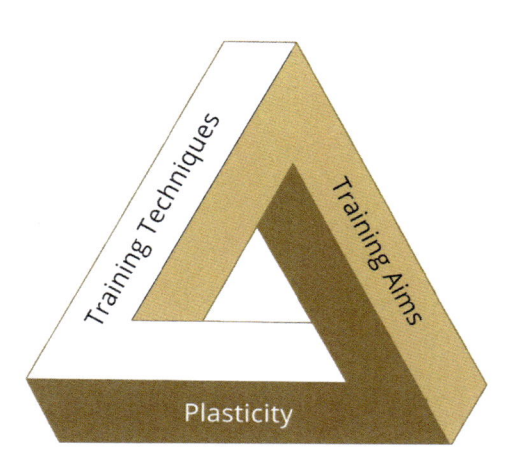

12 Myofascial Training Techniques
Techniques that support fascial plasticity:

1. Stabilising
2. Toning
3. Pushing, Pulling, Counter-traction
4. Active Lengthening and **Expansion**
5. Spiralling
6. Hydrating
7. Gliding
8. Domino Motion
9. Bouncing and Swinging
10. Massaging
11. Active Ease
12. **Melting and Invigorating**

12 Training Aims
Four top achievements of pliability:

1. Dynamic Stability
2. Multidimensional Strength
3. **Limberness**
4. Movement Rhythm
5. Elasticity
6. **Tissue Nourishment**
7. **Adaptability**
8. Resilience
9. Somatic Trust
10. Movement Courage
11. **Mañana Competence**
12. Kinaesthetic Intelligence

IN A NUTSHELL
THINGS WORTH KNOWING ABOUT FASCIAL PLASTICITY AND MOVEMENT

4 Things About Fascial Plasticity

Natural plasticity:	Fascia has natural plasticity.
Forming and pliable:	Fascial plasticity refers to the tissue's ability to form and become pliable to take on a new shape under certain conditions.
Form-giving:	Plasticity gives form to individual structures, such as muscles, and contributes to the contours of the body as a whole.
Shape-changing:	Plasticity may enable fascia to undergo long-term changes in length or alter its consistency temporarily, to reshape immediately with long-lasting effects.

4 Things About Melting

How melting may work:	One explanation is an easeful loop between the nervous, muscular, and fascial system, in which the increased parasympathetic nervous system activity lowers muscle tonus, both of which might facilitate fascial pliability.
Natural variability:	In a healthy body, not all fascial structures are meant to undergo plasticity changes with the same readiness, if at all.
Sense of melting:	The experienced sense of melting may or may not be due to actual length changes in fascia.
Win-win:	Structural change and the felt sense of melting are in different ways equally beneficial.

4 Slings Training Principles

Melting poses:	Melting poses gradually lengthen muscles and fascia in a restful and soothing manner. Generally, the time frame is around one to five minutes.
Melting massages:	Melting massages are either easeful and slow or comfortably sustained. Time frames vary greatly, ranging from ten to twenty seconds, up to a few minutes.
Invigorating motions:	Invigorating motions comprise gentle, yet dynamic movements that facilitate a healthy degree of responsiveness in previously lengthened muscles and fascia.
Melting motions:	Melting motions enhance glide in a slow, generous, and pleasurable manner.

3 BENEFITS OF FASCIAL PLASTICITY

1. Defined body shape

2. Ability to change movement range more long-lastingly

3. Tissue rejuvenation

PART 2

The **tone regulation** of fascia modulates your stability, limberness, and the way you feel.

10. TONE REGULATION

Fascia self-regulates its tone or degree of firmness. In this way, it supports the dynamic balance between stability and pliability based on functional needs, as well as the state of excitement or relaxation in the body.

Let Me Ask You

What makes some people naturally more stable and others more limber regardless of their muscular tone? Why do some inactive bodies become stiffer, while others seem to fall apart? Why are muscle-focussed stretching or core exercises sometimes not enough to attain and sustain lasting flexibility or stability? An important part of the answer to these questions is fascial tone regulation.

Fascia contains contractile cells that can increase or decrease fascial tone, making the tissue either stronger or more pliable. Amongst other factors, movement and how we feel influence the behaviour of these cells.

A Good Reason To Care

If you have tried various methods to improve your flexibility or stability without satisfactory or long-lasting success, then you have good reason to get onto the mat and care for the balance of your fascial tone in a different way.

PART 2

FASCIAL TONE

Fascial tone is a degree of tissue firmness that supports structural integrity by providing dynamic stability for joints and muscles. The latter might sound a bit unusual, but it will make more sense after considering how fascia supports the alignment of bones when standing and moving. If fascial and muscular tone match the functional requirements of the body, joints are stabilised in an adaptable manner. This is in contrast to the limitations in range of movement caused by tissue stiffness and the instability that can result from tissue laxity. As for muscles, if their fascial envelopes and extensions are sufficiently toned, muscles are well-shaped and well-supported in their functions. Fascial armouring, on the other hand, may inhibit healthy muscle activation or relaxation, while tissue laxness requires additional muscular activity, resulting in energy expenditure to make up for the insufficient fascial tone.

Tissue firmness that dynamically stabilises joints and muscles

The question now is how to support healthy tone in a system that cannot be wilfully contracted and relaxed like muscles. To provide practical solutions for this intricate question, we need to speak about the contractile cells mainly responsible for fascial tone regulation, the specialised myofibroblasts.

Largely regulated by contractile myofibroblast cells

FIBROBLASTS: THE BUILDERS, REMOVALISTS, AND FIRST AIDERS

The copious fibroblast cells in fascia have a variety of functions, the main one being tissue remodelling in healthy physiological conditions and during injury. With that in mind, envision the contractile fibroblasts as the builders, removalists, and first aiders of the fascial system.

Building, breaking down, and repair of fascia

Their Roles

Builder: As builders, they lay down the fibrous building blocks of the fascial architecture.

Removalist: As removalists, they break down and clear away decrepit elements.

First aider: As first aiders, they assist tissue repair.

PART 2

Their Actions

Construction: Fibroblasts secrete collagen and elastin proteins for the construction of the fibrous network.

Removal: They also secrete precursors for the enzymes that break down the fibres.

Repair: When tissues are injured, fibroblasts generate traction in the immediate vicinity of the wound to support healing.

Stimulation

Movement stimulates fibroblasts to do their jobs

Fibroblast cells respond to compressive and tensional forces. When pressure is applied, or the tissue is lengthened, these cells are stimulated to more actively remodel the fascial network. Fibroblasts need sufficient movement stimuli to function well – or function at all.

Builders

Karin and Martina building things …

Removalists

... removing things ...

First Aiders

... and mending things at Fremantle Harbour in Western Australia

PART 2

MYOFIBROBLASTS: THE ACTION HEROES

Repair and reinforcement of fascia

The highly contractile myofibroblasts are activated repair cells whose major function is remodelling of fascia, though in a more reinforcing manner than fibroblasts. You can envision these cells as the action heroes of the fascial system.

Their Main Roles

Healers: As healers, the myofibroblasts play a critical role in wound closure.

Reinforcers: As reinforcers, they strengthen fascia by increasing the firmness of the tissue.

Karin and Martina feeling like Superwoman and Super Mario at Fremantle Harbour in Western Australia

Their Superpowers

Repair: Myofibroblasts aid the repair of fascia by pulling the tissue margins together and laying down extra collagen to reknit the wound. Their superpowers are needed to restore tissue integrity after injury.

Reinforcement: When myofibroblasts contract, they pull surrounding tissue tighter. In this way, they locally enhance tone and tension. With the extra collagen they lay down, in time, the fibre density follows suit.

Involuntary Contractility

From the 'Fascia in a Nutshell' chapter, you might remember that myofibroblast contractions are involuntary, therefore beyond our conscious control, and occur very slowly. These cells indicate that fascia contracts and releases over minutes or hours, independent of muscle tonus changes. Although individual contractions are relatively weak, the cumulative effect can significantly increase tissue tone and remodelling.

Involuntary, very slow contractions

Wound Healing and Pathology

„For optimal tissue repair keep your myofibroblasts in balance".   Boris Hinz

While the mechanisms of wound healing are not yet fully understood, it has become clear that both fibroblasts and myofibroblasts play a critical role in the process. The traction forces of fibroblasts and contraction of myofibroblasts are believed to be especially responsible for wound closure.

Wound Healing

After a tissue injury, fibroblasts are stimulated mechanically by altered activation patterns. They are also stimulated chemically by inflammatory mediators. Either way, fibroblasts undergo differentiation into contractile myofibroblasts. The myofibroblasts aid the tissue repair by pulling the tissue margins together and laying down extra collagen, reknitting the wound, and forming a scar.

Myofibroblast activity is crucial for wound healing

Pathology

When the body's self-healing mechanism fails, it either leads to excessive repair or deficient wound healing. Naturally, when a wound closes, myofibroblasts disappear. Too many myofibroblasts working for too long cause excessive shortening and thickening of the tissue. This can lead to pathological contractures and fibrosis, as seen in Dupuytren's contracture (Viking's disease) and hypertrophic scars (a raised scar with excessive amounts of collagen). On the other hand, insufficient myofibroblast activation and activity hinder healing, which leads to chronic wounds.

Excessive or insufficient myofibroblast activity leads to pathology

Healthy and Strong Fascia

Myofibroblasts are present in healthy fascia

Myofibroblasts are often associated with wound healing or pathological contractures. However, these cells are also present in healthy fascia all around the body, though with significant density variations. In individual structures or body regions (and presumably people) where the myofibroblast population is increased, the fascia is firmer. This extra firmness provides greater stability, or when excessive, creates stiffness.

Pointing to a genetic predisposition and fascial tone regulation through movement

The presence of myofibroblasts in healthy fascia might also point towards a genetic predisposition to a higher or lower myofibroblast ratio in the body and the possibility of fascial tone regulation through movement.

FASCIAL VIKINGS AND TEMPLE DANCERS

Working hypothesis with practical value

Based on a working hypothesis by Robert Schleip, we have a genetic predisposition towards higher or lower fascial tone. To use his wonderful, practical, and useful analogy, all of us are somewhere along a spectrum of fascial tone. We range from being a highly stable Viking to being a super bendy Temple Dancer. This implies that someone with innately firm Viking fascia has a higher ratio of myofibroblast cells and is therefore naturally more stable than a limber Temple Dancer with naturally more pliable fascia.

Broad Spectrum of Inherent Fascial Tone

Naturally Stable Viking
On one side of the spectrum is the Viking, who has comparatively high fascial tone, thus greater fascial stability (to various degrees).

Naturally Limber Temple Dancer
On the other side is the Temple Dancer, who has comparatively low fascial tone, thus greater fascial pliability (to various degrees).

Stiff Viking — Stable Viking — Limber Viking — Stable Dancer — Limber Dancer — Unstable Dancer

Who is Who?
Don't judge the book, or a Fascial Viking or Temple Dancer by her or his cover...

Fascial Viking Karin and less-of-a-Viking Martina with a bit-of-Temple-Dancer Luke and Temple Dancer Grant

Who Are You?
So, who are you? Are you more of a fascial Viking or a Temple Dancer?

Before answering this question, pause and introspect, because muscle tone and fascial tone are woven together in our sensory perception, and internal differentiation can be challenging. Only because a person feels restricted and tight doesn't mean they are a fascial Viking, and wanting to see yourself as a limber body doesn't make you a fascial Temple Dancer. In other words, the way we feel or view ourselves might be different from the inherent tone of our fascia. And it might be vastly different from the strength, looseness, weakness, or tightness in our muscles! Some of us Vikings want to be Temple Dancers so badly that we muscle our way through even the gentlest movement class, Viking-style. Meanwhile, the core muscles of overly loose Temple Dancers can be so tight that they keep stretching what is barely holding them together, with detrimental effects for the structural integrity of the body.

Muscle tone, the exterior view and self-perception can vastly differ from fascial tone

Experience has shown that whoever you are and wherever you are on the fascial Viking-Temple Dancer spectrum, awareness and a complementing movement practice actively contribute a degree of fascial tone that facilitates postural ease and movement freedom for your uniquely composed body. The fascia-focussed movement diversity of Slings has proven a success for fascial Vikings and Temple Dancers alike, promoting more suppleness in the naturally stable, while enhancing dynamic stability in the naturally limber.

Understanding the inherent fascial type is key to rebalance fascial tone

PART 2

FACTORS INFLUENCING FASCIAL TONE

A wide variety of factors influences fascial tone. For example, substances released during inflammation and wound healing, and alterations in the pH level due to nutrition or chronic over-breathing, can lead to fascial contractions.

Inflammation, nutrition, and breathing habits

From shared experiences and observations, movement behaviour seems to play a crucial role in fascial tone regulation. A highly active person who regularly loads their fascia with intense training tends to have firmer fascia than someone opting for a gentler or more introspective form of exercise. Note, this is often true for Vikings, possibly due to their genetic predisposition. However, Temple Dancers, on a mission to tone their fascia Viking style, may not always attain the desired result by simply working hard. Fascial tone regulation is multifaceted. Try adding some extra deep breaths and a double dose of parasympathetically sponsored mañana competence to the workout mix.

Activity level and training intensity

Speaking of the nervous system, another significant influence on fascial tone regulation is the dynamic balance in the autonomic nervous system. It is suggested that the sympathetic nervous system, which excites and alerts us, can trigger fascial contractions. These contractions could indicate that the stiffness some of us perceive when being or feeling deeply stressed over an extended period of time, might be fascial, rather than muscular. Hence the short-term pleasurable effect of a stretch or a massage.

Dynamic balance in the autonomic nervous system

PART 2

The sympathetic nervous system (SNS) activates
The parasympathetic nervous system (PNS) rejuvenates

Dialogue with the Autonomic Nervous System

The autonomic nervous system (ANS) can be divided into the sympathetic and parasympathetic nervous systems. The sympathetic division regulates the use of metabolic resources and the coordination of the body's emergency response (fight or flight). The parasympathetic division usually governs the restoration of metabolic reserves and the elimination of waste products (rest and digest).

Intense and/or sustained sympathetic nervous system (SNS) activity is believed to increase myofibroblast activity, therefore enhance fascial stiffness, at least in part. Our working hypothesis is that by regularly enabling parasympathetic nervous system (PNS) activity and subsequently quieting the sympathetic branch, healthy fascial tone regulation can be supported proactively, at least to a degree. Taking this a step further, we work from the premise that whatever occurs between the autonomic nervous system and the fascial system is causative, not correlative. Correlation states that the systems are interdependent and have reciprocal relations in which one system affects the other. Causation would mean that one system creates an effect in the other while remaining unaffected by the change it created. In that sense, well-regulated fascial tone positively influences the way we feel, which aids dynamic balance between the sympathetic and parasympathetic divisions of the autonomic nervous system.

Sustained SNS activity can stiffen fascia
Regular PNS activity can assist healthy fascial tone regulation

Sustained Feeling of Distress

Stiffening Cycle
(Sustained SNS activity → Increased fascial stiffness → Feeling 'overactivated' → Sensory 'noise' →)

Dynamic Balance between Systems

Healthy Self-Regulation Cycle
(Dynamic balance between the SNS and PNS → Healthy fascial tone regulation → Feeling invigorated and rejuvenated → sensory 'quiet' →)

Muscle Tightness or Fascial Stiffness?

Most of us have experienced hardening effects when intensely distressed or feeling taxed over an extended period of time, at least in some parts of our bodies. Until fascia came into focus, the tightness was primarily associated with muscles. However, with what we know now about the response of fascia to sustained sympathetic nervous system activity, it is probable that fascia is just as affected. While short-term distress can lead to muscle tightness, long-term distress might create fascial stiffness.

Short-term distress: muscle tightness
Long-term distress: fascial stiffness

Being and Feeling Armoured or at Ease

The effects go beyond tissue stiffening. A fascial 'armour' changes how we feel, which changes our posture and the way we move or no longer move. Nervous system activity, fascial tone, and how they make us feel, all affect each other. This can create either a positive or a negative feedback loop. Naturally, the movement aim is to sustain a positive feedback loop or to gradually turn a negative into a positive one.

Structurally armoured and feeling held (back)

Structurally supported and feeling at ease

Structural & Sensory Feedback Loop

(Influences) Fascial tone regulation → (Influences) How we feel → (Influences) How we move or no longer move →

SLINGS IN MOTION

FASCIAL TONE REGULATION THROUGH MOVEMENT

Now we arrive at the exciting and quietly invigorating part of this chapter: fascial tone regulation through movement. Because pathological conditions and nutrition are outside the scope of this book, the suggested training strategies focus on promoting tissue nourishment directly and support a positive feedback loop between the nervous and fascial system.

Tissue nourishment and dynamic balance in the ANS

The practical applications for fascial tone regulation are derived from the previously discussed information and supported by experience. Included are movement strategies that facilitate fluid flow and glide to nourish strained fascia locally, at a tissue level. We also take deep breaths, 'unwind' the spine, and practise 'mañana competence' to foster healthy activation of the parasympathetic division of the autonomic nervous system, while quieting down excessive noise in the sympathetic division.

Well-Toned and At Ease Not Loose and Unstable
Note, that the proposed movement strategies supporting parasympathetic nervous system activation don't mean or intend to down-regulate fascial tone to a state of laxity. Feeling at ease doesn't mean your fascia goes slack. In fact, in practice, we sometimes see the opposite. 'Overactivated' bodies that are structurally too loose for their own good in many places, while holding excessive tension in areas deep inside (core tension), gradually develop a more balanced fascial tone, inside and outside.

Fluid Flow

Hydration and glide for tissue health

To support fascial tone regulation locally, we include movement strategies that promote tissue hydration and glide. The reasoning is as follows:

- Exercises that enhance hydration and glide facilitate fluidity and fluid flow fascia.
- Fluid flow in fascia is vital for tissue health.
- Tissue health is essential for optimally regulated fascial tone.

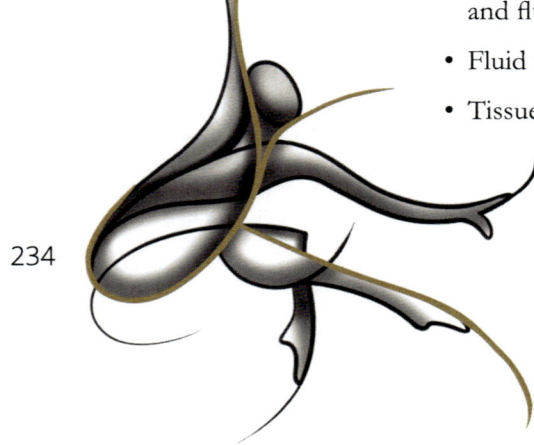

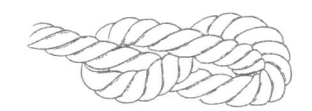

Positive Effects on the Tissue
Tissue hydration and glide can aid fascial tone regulation at a tissue level, which then creates a positive movement cycle.

Movement ease:	Fluid and well-toned fascia facilitate movement ease. Moving with ease facilitates fluid flow and optimal tone in fascia.

Positive Effects on Feelings
Well-hydrated and gliding fascia positively influences the way we feel in our body and with our movements. In this way, another health-promoting cycle is set into motion.

Sense of wellbeing:	Healthy fascia feels good, and feeling good supports the dynamic balance between the sympathetic and parasympathetic divisions of the autonomic nervous system. In turn, this benefits fascial tone regulation.
Movement motivation:	Healthy fascia facilitates movement ease, which motivates us to keep moving. In turn, moving regularly promotes fluidity and a healthy tone in fascia.

Slow, Deep Breath

Before reading on, pause, and take a conscious, full breath. Inhale deeply and exhale slowly. Volumes have been written about the health-promoting and healing qualities of the breath, and for good reason. Many ancient and contemporary movement and meditation practices have the breath at their centre. Taking relaxed, deep breaths is natural medicine that activates the parasympathetic nervous system, thereby inducing muscle relaxation and assisting fascial tone regulation. So, before turning the page, take five more nourishing breaths. And to enhance the healthful effects, lift the corners of your lips too: seriously!

Slow, deep breaths for PNS activation

Unwinding the Spine

Most of us have experienced how different spinal movements, or inhibition of those movements, can greatly affect the way we feel, physically and emotionally. This is not surprising as the spine houses the spinal cord. Together with the brain, the spinal cord forms the central nervous system, which extends to the peripheral nervous system. The peripheral nervous system includes the somatic nervous system and the autonomic nervous system. While the nerves of the somatic nervous system serve the skeletal muscles and skin, the autonomic nervous system regulates homeostatic functions.

Spinal movement can have a more stimulating or calming effect

Simply speaking, activity in the somatic nervous system, in collaboration with the muscles surrounding the spine, modulates the way we move our back. The movements are shaped and changed with the functionality of the fascia, which – as previously discussed – is in close contact with the autonomic nervous system. While some activities tend to stimulate the sympathetic branch of the ANS, others facilitate a more parasympathetic response. Which is stimulated depends on a multitude of factors, not only the type of motion we perform. Influencing factors include the range, dynamics, duration, and intensity of a movement, if the movement is done actively or passively, breathing patterns, the pre-existing balance in the ANS, a person's perception of the immediate environment, and their past experiences. Also, when utilised, the size, consistency, surface, and placement of props can make a noticeable difference. For example, a spine extension tends to have a more stimulating, sympathetic effect. However, if it is executed passively in a supine position with two relatively small, soft massage balls in the mid-thoracic region, combined with deep, expansive breaths, it can have a wonderful melting quality, feeling deeply restorative.

With that in mind, treat the following experience-based listings as a generic guide of movements that can induce either a positively stimulating sympathetic or calming parasympathetic response.

Sympathetically-Oriented Spinal Motions

Extension:
- Sustained pose, especially with lumbar extension included
- Segmental movement against gravity

Spiralling:
- Segmental movement, especially when dynamic

Parasympathetically-Oriented Spinal Motions

Flexion:
- Restful passively sustained pose
- Easeful actively sustained pose
- Segmental movement, more so when working with rather than against gravity

Rotation:
- Restful passively sustained pose

Lateral flexion alone or combined with spiralling:
- Soothing, generous motion

Easeful curl, restful twist, and soothing side bend with spiralling for inner calm and restoration

Spiralling Mermaid with Slow Domino Motion

Parasympathetically-Oriented Exercise Examples

Neck Massage & Nodding

Easy Twist

PART 2

Die Mañana-Kompetenz: Entspannung als Schlüssel zum Erfolg

Mañana Competence

A special thank you to Maya Storch and Gunter Frank, the authors of "Die Mañana-Kompetenz: Entspannung als Schlüssel zum Erfolg" for their permission to include the term they coined in the Slings concept!

Every day, we do the best we can to make the most of what is. In our own ways and within our own capacities, we juggle the many things that make up our lives. We streamline our schedules and upgrade our devices to create more efficiency. Yoga has become a billion-dollar industry, mindfulness retreats can be done in the woods or five-star wellness centre. Depending on your requirements, budget and timeframe an abundance of organisations, individuals, books and online lectures can be found to apparently improve our work-life balance, inner balance, or outer beauty. Though sometimes before we know it, the contemplative movement lesson, weekend meditation retreat, or dinner with friends have become an additional task on the to-do-list. Life is busy and full of activities, yet might feel empty in other, inner ways … perhaps because there is a healthy dose of mañana competence missing!

Mañana What?

Ability to take it easy and let the PNS take over

The expression mañana competence was coined by two insightful doctors. Gunter Frank and Maja Storch believe that sustained health is not a matter of better planning, but of getting (back) in touch with how we feel and what we truly need. Mañana is Spanish for 'tomorrow'. In our context, mañana competence stands for the ability to take it easy and let the parasympathetic nervous system take over; in motion and stillness.

Parasympathetic Boost

Recovery and rejuvenation

While the sympathetic nervous system motivates us for action, the parasympathetic nervous system boosts recovery and rejuvenation. Activation of the parasympathetic division is directly linked to a sense of inner peace, sensuality, and being present. It also makes you a kinder and better-looking person in your own eyes and the eyes of others.

Are You Mañana Competent?

For those of us who are experts at hurrying through life and juggling multiple projects at once, practising mañana competence is highly recommended! Are you wondering if this concerns you too or how mañana competent you are? The following are qualities that signify mañana competence.

- You are present.
- You feel your body.
- You are aware of your emotional states.
- You sense your environment.
- You have empathy for yourself and others.
- You are physically and emotionally resourceful.
- You adapt readily.
- It feels enough to just be.
- You have zest for life.
- You feel vibrantly alive (instead of surviving life).

Being present, in tune, resourceful, and contented

PART 2

SLINGS FORMULA FOR PRACTISING MAÑANA COMPETENCE

"For me, mañana competence has come to mean a carrying of the calm and grounding of my practice into tomorrow – no matter what that tomorrow brings."
Heidi Savage

Engagement with presence is key

Both stillness and contemplative movement are beautiful ways to practise mañana competence. Either way, the key is to not slip into a mindset of 'getting it done'. Whatever practice you engage in, ensure that you are present, and feel physically and emotionally nourished by what you do. While ticking a class off the to-do-list or absentmindedly going through the motions might still benefit the body on some level, it doesn't give you all it could, nor does it strengthen your mañana competence.

To assist in practising mañana competence in motion, Slings offers you a flexible 'formula' or lesson structure. It aims to facilitate engagement, and with it presence, winding down physically, mentally, and emotionally, as well as rest, regeneration, and invigoration of body and mind. Ideally, the practice leaves you in a state of calm vitality.

1. State of Flow Through Contrast

Engagement is a result of presence and sustained interest in an activity. When the activity is too easy, interest dwindles. When it is too hard, enjoyment wanes. Sustained engagement and the sense of achievement that comes with it require a balance between ease and challenge.

Balance between ease and challenge

In Slings, practising mañana competence is not done with passive poses or relaxation exercises per se, although they can be part of it. Instead, we aim to facilitate a state of flow with contrasting movement sequences that engage and satisfy body and mind alike.

Contrasting movement sequences to sustain engagement

2. Gentle Flow to Wind Down

To deepen the delicious feeling of slowing down, mellow and gently flowing exercises are woven into the class toward the end of the practice. The calmness of slow movements can have beautifully soothing and restorative effects that exceed the duration of the lesson by far.

Mellow motions to slow down and restore

3. Absorption

At the end of a lesson, a time of absorption is of tremendous value for body and mind. 'Absorption' is a state of physical stillness, mental presence, and emotional awareness. It is a time of resting within that facilitates the integration of what has been experienced during the lesson.

Physical stillness, mental presence, emotional awareness

Ideally, the ambiance in the room supports the process. This may be through silence, by gently 'anchoring' thoughts in the present moment, or a combination of the two. For some of us, silence is a deeply nourishing respite from the noise pollution we are immersed in every day. For others, the absence of distracting sounds leads to a mental escape in which thoughts distract from feeling current physical sensations and affective emotions. A guided body scan, breath awareness, listening to the lyrics of a wholesome song, or the words of a meaningful story can beautifully 'anchor' the awareness in the present moment. The feeling of 'letting go' that regularly accompanies this quiet activity can be a wonderful passageway to embracing silence.

Resting within to integrate experiences

PART 2

Gentle activity to awaken the senses

4. Invigoration

Absorption is complemented by invigorating movement. 'Invigoration' comprises gentle activity that wakes up the senses, so you feel physically and mentally even more refreshed. It also builds a bridge from the serenity on the mat to the demands of daily life, assisting the integration of practised mañana competence into your very being.

Calm Vitality

Supportive of fascial tone regulation and dynamic balance in the ANS

At its best, the Slings practice leaves you with a sense of calm vitality, a state that supports healthy fascial tone regulation and dynamic balance in the autonomic nervous system.

SynerChi Essential Flow
Arm Arc & Dynamic Plié with Arm Circle

Roll Down & Forward Fold Leg Stretch

Active
Forward Fold

Long Lunge

Dynamic Long Lunge into Inverted V

Big Wave into Front Support

Kneeling Front Support

Small Wave

Kneeling Upward Stretch & Active Child's Pose

Front Support

Inverted V

Forward Fold Leg Stretch & Rolling Up

Repeating sequencing until Active Child's Pose

PART 2

Child's Pose   Low Kneeling into High Kneeling

Side Sit   *Transition:* Z-Sit   Basic Roll Up Preparation

Spiralling Twist & Curl in Basic Long Sit

Transition: Basic Rolling Down into Rest Position

Pelvic Curl

244

Fish Pose with Shoulder Massage

Fish Pose Melting

Absorption

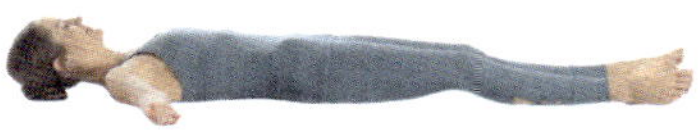

Invigoration:
Long Stretch

Rest Position

Rocking into Rolling

Rolling Like a Ball into Sitting

Spiralling Twist

Arch    Curl

ONTO THE MAT AND INTO THE BODY

▶ Story and Practice: Tone Regualation in Motion

SLINGS MYOFASCIAL TRAINING TECHNIQUES AND AIMS

The following Slings Myofascial Training Techniques and Training Aims shown in gold, bold font are directly related to fascial tone regulation.

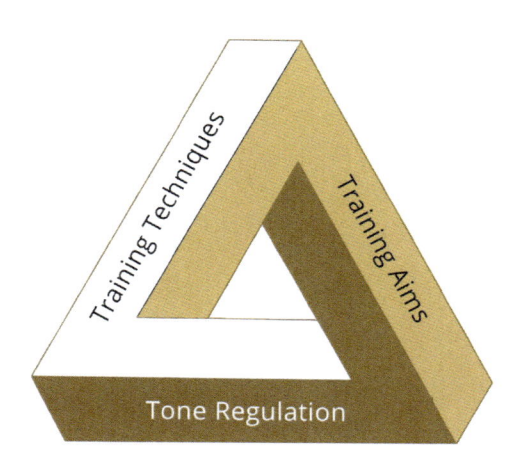

12 Myofascial Training Techniques

Techniques that directly influence fascial tone:

1. Stabilising
2. Toning
3. Pushing, Pulling, Counter-traction
4. Active Lengthening and Expansion
5. Spiralling
6. **Hydrating**
7. **Gliding**
8. Domino Motion
9. Bouncing and Swinging
10. Massaging
11. **Active Ease**
12. **Melting and Invigorating**

12 Training Aims

Four top achievements of tensegral balance:

1. Dynamic Stability
2. Multidimensional Strength
3. Limberness
4. Movement Rhythm
5. Elasticity
6. Tissue Nourishment
7. Adaptability
8. **Resilience**
9. **Somatic Trust**
10. Movement Courage
11. **Mañana Competence**
12. **Kinaesthetic Intelligence**

IN A NUTSHELL
THINGS WORTH KNOWING ABOUT FASCIAL TONE REGULATION AND MOVEMENT

4 Things About Fascial Tone Regulation

Fascial tone:	Fascial tone is a degree of tissue firmness that supports structural integrity by providing dynamic stability for joints and muscles.
Fascial tone regulation:	Fascial tone is largely regulated by the activity of fibroblast and myofibroblast cells.
Viking or Temple Dancer:	There might be a genetic predisposition towards a higher or lower fascial tone. It places all of us somewhere along a spectrum that ranges from highly stable Viking fascia to super limber Temple Dancer fascia.
Influencing factors:	Fascial tone is influenced by a wide variety of factors such as tissue health, nutrition, breathing, and movement behaviours, as well as activation patterns of the autonomic nervous system.

4 Things About Fibroblasts and Myofibroblasts

Fibroblast functions:	The abundant fibroblast cells are the 'builders', 'removalists', and 'first aiders' of the fascial system. The building, breaking down, and repair of fascia.
Myofibroblasts functions:	The highly contractile myofibroblast cells are the 'action heroes' of the fascial system. They repair and reinforce fascia.
Myofibroblast contractions:	Contractions are involuntary and very slow. Although individual contractions are relatively weak, the cumulative effect can significantly increase tissue tone and remodelling
Healthy fascia:	Myofibroblasts are present in healthy fascia all around the body, although with significant density variations. This points to a genetic predisposition to higher or lower fascial tone and the possibility of regulation through movement.

4 Things About the Autonomic Nervous System

Autonomic nervous system:	The sympathetic nervous system motivates action, while the parasympathetic nervous system facilitates recovery and rejuvenation.
Sympathetic activity:	Sustained sympathetic nervous system activity can stimulate myofibroblast activity, thereby increase fascial stiffness.
Parasympathetic activity:	Regular parasympathetic nervous system activity can aid fascial tone regulation by quieting the sympathetic division and with its regenerative power.
ANS-fascia dialogue:	Dynamic balance in the autonomic nervous system (ANS) supports healthy fascial tone regulation. In turn, well-regulated fascial tone has a healthy feel, which supports dynamic balance in the ANS.

5 Slings Training Principles

Fluid flow:	Movement strategies that promote hydration and glide nourish fascia, therefore support fascial tone regulation locally, at the tissue level.
Deep breath:	Relaxed, slow breathing is natural medicine that activates the parasympathetic nervous system and thereby assists fascial tone regulation.
Easeful spinal motion:	Restful poses in which the spine is flexed or rotated, as well as soothing motions that feature spine flexion or lateral flexion in combination with rotation, can have a calming, restorative effect.
Mañana competence:	Mañana competence is the ability to take it easy and let the parasympathetic nervous system take over. With it comes presence, connectedness, resourcefulness, and contentment.
Calm vitality:	The Slings 'formula' for mañana competence, and therefore a sense of calm vitality, encompasses: 1. State of flow through contrast, 2. Gentle flow to wind down, 3. Absorption, 4. Invigoration.

3 BENEFITS OF FASCIAL TONE REGULATION

1. Structural integrity
2. Fascial repair, rejuvenation, and reinforcement
3. Calming inner sense of togetherness

PART 2

The **kinaesthesia** of your fascia gives you self-awareness,
enables well-orchestrated movement,
and informs you of how you feel about the way you feel
to motivate healthy body choices.

11. KINAESTHESIA

Fascia is the most influential organ of kinaesthesia and the richest sensory system in the body. It is a major contributor to kinaesthetic intelligence, the synergy of proprioceptive finesse, and interoceptive clarity.

Let Me Ask You

Where does movement coordination come from, or seemingly elusive experiences like physical spaciousness, heaviness, or buoyancy? Why can movement make you feel happy or irritated? And how do you know that every part of your body belongs to you? Because of your kinaesthetic sense!

Through our eyes, ears, nose, mouth, and skin, we receive information from the world around us. Through our fascia, we receive information from the world within us. The fascial system provides us with a sixth sense that enables us to feel the body, coordinate and sense the quality of our postural alignment and movements, and be aware of emotional states. It also motivates us to make health-oriented behaviour adaptations. Supernatural achievements, right? That is not all, though. Do we ever wake up in the morning wondering who we are? Or wondering whose hand is holding our coffee cup? Of course not! Kinaesthesia gives us a complete inner image of ourselves, which manifests in physical self-awareness and a feeling of being at home in the body.

A Good Reason To Care

The mere ability to stand and go about your daily business depends on your kinaesthesia. Add postural and movement ease, an up-front body language, emotional clarity, and healthy choice-making, and you have many compelling reasons to get onto the mat and train your kinaesthetic intelligence regularly.

THE FEELING OF SENSATION AND EMOTION

When sensations become emotions, vivid feelings emerge.

Sensation and emotion (e)merge to feeling

Before taking a detailed look at kinaesthesia and what kind of feelings it encompasses, we need to have a mutual understanding of what it means to feel. So, at least for the duration of this chapter, let's agree that feeling in the context of kinaesthesia encompasses both sensation and emotion. Let's also agree that sensation and emotion influence each other.

Sensations tell you what you feel

Sensation: A sensation is the perception of something physical. A physical sensation creates an emotional response.

Emotions tell you how you feel

Emotion: An emotion tells you how you feel about a sensation. The emotional state changes how the physical sensation is experienced.

Here is an example to make this more tangible.

Imagine you are doing a self-massage exercise in which you are rolling the sole of your foot over a spiky ball. Feeling the ball touch the skin at the bottom of your foot is a physical sensation. The sensation creates an emotional response, for example, pleasure or irritation. Being pleased or irritated by the sensation informs you, at this moment in time, how you feel about it. Maybe tomorrow, your emotional response will be something else. Or if you stepped on the ball accidentally, you would feel differently again. Experiencing pleasure instead of irritation changes the quality, and generally also the intensity with which the physical sensation is perceived. Emotions tell you what you want more of, less of, or what you don't want at all.

Lucas and Karin feeling off centre and amused about it in Berne in Switzerland

PART 2

KINAESTHETIC SENSE

"Our central nervous system receives its greatest amount of sensory input from myofascial tissues."
Robert Schleip

Movement sense encompassing proprioception and interoception

Kinaesthesia means movement sense. It can be subdivided into two categories:

Proprioception: The proprioceptive sense enables unconscious and conscious coordination of body alignment and movement.

Interoception: The interoceptive sense facilitates the unconscious and conscious ability to feel the quality of physical sensations, their effect on emotional states, and the ability to adapt in a wellbeing-oriented manner.

While proprioception has always been directly associated with kinaesthesia, interoception was commonly linked to viscera and physiological balance (homeostasis). In the last few years, though, the meaning of interoception has shifted. The restrictive view of interoceptive sense solely stemming from the internal organs, gave way to the more inclusive view of using interoception as an umbrella term for the phenomenological (felt) experiences of the body's states.

Brainy Difference

The difference between proprioception and interoception doesn't lie in their importance for physical functionality. Instead the difference lies in the brain, where different areas process proprioceptive and interoceptive information.

Sensorimotor cortex: Proprioceptive information is processed in the sensorimotor cortex of the brain.
Voluntary movement is coordinated here.

Insular cortex: The insular cortex or insula of the brain is largely responsible for processing interoceptive information.
Activity in this region is associated with the experience of emotional states.

FASCIA: ORGAN OF KINAESTHESIA

If the brain is the chief organ of cognitive intelligence, then fascia is the principle organ of kinaesthetic intelligence.

If the fascial system could be unravelled, its surface area would by far surpass our largest organ, the skin. Fascia is not only vast in terms of its spatial occupation, it is also in a league of its own as a sensory system.

Fascia is densely populated with sensory receptors. Compared to the number of muscle spindles, there are many more mechanoreceptors in fascia than in muscle tissue. Including tiny free nerve endings in the equation, the number of sensory receptors in fascia is even superior to those in the retina of the eye. In this sense, fascia can be viewed as the body's most influential perceptual system and organ of kinaesthesia.

High density of sensory receptors

Influential sensory system

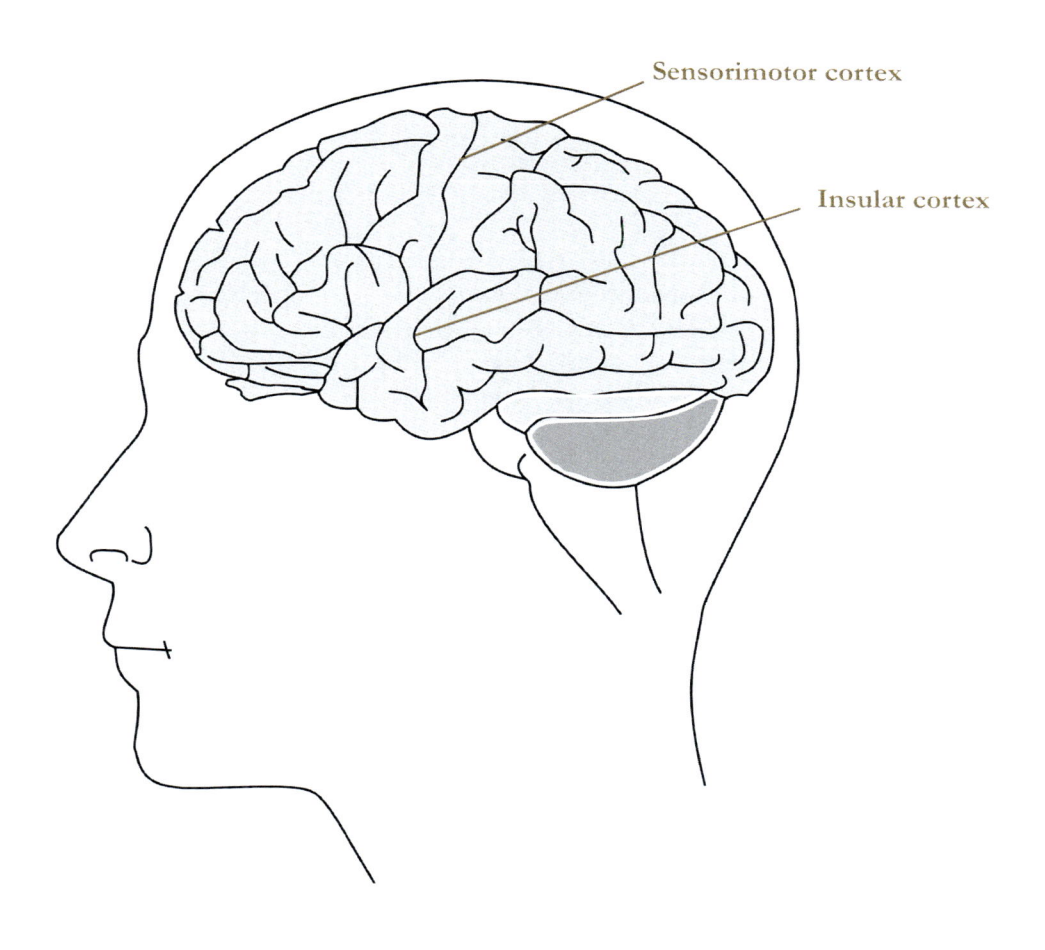

Sensorimotor cortex

Insular cortex

Stimulation through tension, pressure, shear, vibration

Fascial Mechanoreceptor Quartet

Fascia contains sensory receptors that respond to mechanical stimuli. The mechanoreceptors in focus are the:

1. Golgi organs
2. Ruffini corpuscles
3. Pacini corpuscles
4. Interstitial receptors

What stimulates them as a group is tension, pressure, shear, and vibration. The first three are the most common types of load in fascia-focussed training. Movement-related examples below, include but are not limited to, stimuli to which individual fascial mechanoreceptors respond.

Golgi: Proprioceptively-oriented

Pacini: Proprioceptively-oriented

Ruffini: Proprioceptively and interoceptively-oriented

Interstitial: Interoceptively-oriented

A Word About the Interstitial Receptors
Of the four mechanoreceptors, the interstitial receptors are the most abundant in fascia. This large group of free nerve endings can be subdivided into two groups: receptors with a low stimulation threshold and receptors with a high stimulation threshold. In essence, interstitial receptors can perceive both feather-light touch and firm pressure. Besides functioning as mechanoreceptors, they interact closely with the autonomic nervous system. Stimulating these receptors through movement or self-massage can positively influence blood flow, respiration, heart rate variability, and other vital functions.

PROPRIOCEPTION: SENSING BODY AND MOVEMENT

Proprioception can be described as the ability to sense structural relationships and how they relate posturally and in motion. The resulting neuro-myo-fascial coordination enables ease and efficiency when standing and moving.

Well-coordinated postural alignment and movement

INTEROCEPTION: EMBODYING EMOTIONS

"The vividness of your interoceptive body image determines how well emotions and cognition are embodied."
Paraphrasing Bud Craig

Interoception can be described as the moment-to-moment process in which the sensations of the body are experienced and neurologically integrated. Through this process, we assess and emotionally respond to what we feel.

Moment-to-moment process in which bodily sensations are felt and emotionally responded to

Without your interoceptive sense, you wouldn't be or feel alive, nor would you recognise yourself as a whole human being. Because interoception serves homeostatic functions, it is directly associated with the maintenance of life-sustaining processes. It also facilitates the perception of emotions by creating a sensory image of the condition of the body. Within this image, the emergent feelings provide a basis for self-awareness. Research from various fields, including the extensive studies of neuroanatomist Bud Craig, have shown that interoception facilitates:

- Physiological self-regulation for maintaining or regaining optimal health and wellbeing

- Efficient energy utilisation in the body

- Health-promoting behaviour adaptations

- The feeling of being alive

- The feeling of physical wholeness

- Recognising all aspects of what it means to be 'me', physically, mentally and emotionally

- Emotional self-awareness

- Awareness of other people's emotional states

- Advantageous choice-making based on subtle environmental cues

- Subjective time perception

Learning the Meaning of Feelings

The body knows what is healthy for you. Interoception is the sense that tells you all about it.

The meaning of feelings is largely learned

Essentially, interoception is about feeling and interpreting feelings to understand what we need to do to stay well. When it comes to the interpretation of feelings, it is valuable to remind ourselves that the current meaning we draw from sensations and emotions is shaped by our past experiences. As children, many of us were taught what the colour blue looks like, and that a blue sky is a good reason to feel happy – happiness by association, so to speak. We learned what hot feels like, and that heat can be dangerous. We were shown that certain emotions and gestures are more appreciated than others, or by others. The emotional element is also represented in our language, where 'gut feelings' and 'cold-heartedness' trigger an immediate emotional response in many of us. Over the years, we have been strengthening these associations, connecting certain physical sensations, body language, and verbal language firmly to certain emotions. When tuning in or asking another person, "How do you feel?" it is good to remember that the answer is only a partial expression of the current state. The remainder is shaped by multifaceted life experiences.

(Re)Uniting Body, Mind, and Emotions

Learned dualism is (re)united by interoception

Staying with the theme of experiences that shape the way we feel, for a moment, let's ponder how learning creates dualism and how interoceptive clarity facilitates (re)union.

As for our movement skills, we have been acquiring and strengthening beliefs about our body-mind relationship throughout life, knowingly or unknowingly. While growing up, we learned to pay attention to our bodies, and we also learned to understand the difference between what the body needs and what our mind, or the mind of another person, wants. Historically, maybe you felt satisfied after eating half of your dinner, yet you were asked to finish your plate. Or you were told to sit still at school when your body wanted to run around outside. And at one point, all of us learned that certain bodily urges require an appropriate place and time to be socially acceptable. In other words, we grew up with an implied understanding that what the body needs and what the mind demands are often different. This apparent conflict is called dualism. As socially appropriate as it might be, it creates a more or less conscious conundrum that raises questions of causation rather than correlation. Interoception (re)unites physical sensations, emotional states, neural integration, and homeostatic functions; therefore, changes the question from: "what causes what?" (causation) to "how do they relate to each other?" (correlation).

KINAESTHETIC INTELLIGENCE

Kinaesthetic intelligence is the synergy of well-orchestrated movement, a vivid whole-body image, and embodied emotions.

Kinaesthetic intelligence represents the level of, and dynamic balance between proprioceptive finesse and interoceptive clarity. However, training these qualities is not simply about being able to do more advanced exercises or feeling more. It is about developing a new degree of movement ease and making sense of what is felt.

To conceptually depict the difference between proprioceptive finesse and interoceptive clarity, I put them in different, yet complementing categories, outlined on the following pages.

*Dynamic balance of **proprioceptive finesse** and **interoceptive clarity***

Karin meets kinaesthetic intelligence and movement freedom on a beach in Zanzibar in Tanzania

PART 2

Ability to coordinate movement in a smooth and well-timed manner

PROPRIOCEPTIVE FINESSE

Being refined in the way you align your body and coordinate movement.

Richard creating a defined shape in Perth in Western Australia

Proprioception

Alignment: The perception of the body's position in space and the relationship of internal structures

Coordination: The ability to orchestrate movement in individual body parts in a coherent way

Rhythm: The skill to move in a well-timed manner that matches the activity

Well-Orchestrated Self-Organisation
Proprioceptive clarity facilitates efficient and poised self-organisation of movement.

Objective Hard Skill
Because proprioception can be described, observed, and assessed, it is considered more of a hard skill. Proprioceptive learning can be supported by clear instructions and mastered with the repetition of deliberately executed movements.

Mover Within
Proprioception is viewed as the mover within because you can't move without it.

Action
Proprioception coordinates your actions.

INTEROCEPTIVE CLARITY

Being clear on how you feel about the way you feel.

Alexa feeling the way she feels with clarity in South Beach in Western Australia

Emotional awareness and the ability to make sense of what is felt

Interoception

Feeling whole: The perception of the body in its entirety

Recognising emotions: Awareness of emotional states

Deriving meaning: The ability to make sense of what is felt on a physical and emotional level

Adapting behaviour: The ability to adapt in a manner that promotes health and wellbeing

Embodied Emotional Awareness
Interoceptive finesse facilitates sensing the body with awareness of emotional states and moving in a wellbeing-oriented manner.

Subjective Soft Skill
Because interoceptive feelings are personal, intangible, and interpretive, it is considered a soft skill. Interoceptive learning takes place in mindful movement that brings awareness to the perception of bodily sensations and emotional responses to what is felt.

Motivator Within
Interoception is the motivator within because the way you feel motivates movement behaviour.

Gesture
Interoception modulates your body language.

PART 2

KINAESTHETIC INTELLIGENCE DYNAMICALLY BALANCED

In summary, the dynamic balance and interplay between proprioceptive finesse and interoceptive clarity characterise the quality of a person's kinaesthetic intelligence.

Proprioception coordinates body alignment and movement	Interoception enables feeling the effects of body alignment and movement, and prompts healthy adaptation
Proprioception can be trained through repetition	Interoception can be trained through awareness
Proprioception moves	Interoception motivates movement
Proprioception organises actions	Interoception modulates body language

EMOTIONAL AWARENESS AND (RE)INTEGRATION OF FEELINGS

Interoceptive Cost-Benefit Movement Decisions

Most of us have, at one point, engaged in movement activities that were not solely in the best interest of our health and wellbeing. If no personal example comes to mind, think of someone playing rugby, engaging in a boxing match, or running an ultra-marathon. All of these activities can fire up a whole array of positive emotions, and training for them undeniably has its physical benefits, yet each of them also extracts a cost. As discussed in the chapter about adaptability, we consciously or unconsciously make cost-benefit decisions based on what we value. It's called being human. Still, now and then, it is worthwhile to assess if the benefits continue to balance or – even better – outweigh the costs. Here is where interoceptive awareness is key. Once a person understands and embodies the difference between sensation and emotion, they can make more informed decisions about what serves them and what doesn't serve them (anymore), both on a physical and emotional level.

Interoception supports choice-making in the body's best interest

Alarmed by Emotions

In our everyday feelings, sensations and emotions naturally blend, which is also true during movement practice. That is fine most of the time. However, there are moments when unawareness of the difference between the two can be a real obstacle to somatic progress and overall wellbeing. For example, a person with back pain history might equate a natural stretch sensation along their spine with the alarming tension they experienced during an acute phase. The fear triggered by the pain anticipation (not pain perception) can lead to movement avoidance. What the person protects themselves from in this scenario is not damage to their body; it is emotional distress. While this response is understandable, the long-term consequences are dire. When difficult emotions are experienced during movement, it is worthwhile to remember that alarming, irritating, or confusing emotions are certainly a sign of something, though not necessarily of imminent, physical danger. A kinaesthetically intelligent way to handle such feelings is with parasympathetically empowered mañana competence, instead of a sympathetically fuelled fight or flight response.

Alarming emotions are a sign of something, though not necessarily of physical danger

INTEROCEPTIVE SOUNDSCAPE

It is now a good time to talk about the feelings we often hold in focus, the ones we sometimes underappreciate, and those that can no longer be perceived. Equating feelings with sounds, we give this next discussion a sensory quality.

Noisy Places

Loud feelings that claim most attention

The loudest 'noises' in the body tend to get the most attention. Pain is blaring, drowning out everything else. Less disconcerting sensations such as muscles hard at work, body parts intensely stretched or firmly massaged let us know – loud and clear – that something is happening inside of us, which can be interpreted as good or bad.

When someone describes how or what they feel in their body, they frequently speak about the noisy places, where they feel strong sensations. A few things are worth remembering at this point:

- Where the noise is felt, may or may not be where it originates. Think of tensegrity, in which an issue in one place can create symptoms elsewhere in the system.

- The noise and its volume are created by physical sensation in concert with an emotional response. It is not one or the other.

- The body is more than its noisiest parts. Bringing awareness to the quietly contented places and attending to silenced areas is also important.

For more ease, soothing and expanded awareness is required

For peace of mind, comfort in the body, and the ability to sense more subtle feelings, it is absolutely useful and often necessary to bring more ease and quiet to noisy places. Attending to symptoms is not only beneficial in the short-term; it might also assist in locating and rebalancing the distant source of the issue.

Quiet Places

This brings us to the body's 'quiet' places. In these areas, we feel somatically at ease. Think of the gentle purring of a gratified cat. Quiet places are subtle, frequently overlooked, or underappreciated. It may be that we take them for granted, or the noise elsewhere absorbs too much of our attention (it's hard to hear the purring of a cat amidst a roaring rock concert). The importance of being aware and appreciating quiet places cannot be overstated. By doing so, we recognise our physical resources, strengthen the trust in the body, and bring calm to a system that might feel dissonant in certain places.

Areas of somatic ease

Appreciation strengthens inner resources and trust

Silenced Places

'Silenced' places are tricky to recognise. We no longer feel them. What adds to the challenge is the common misconception that if a body part can't be felt, it must be fine. The flaw in this is evident when thinking of a paralysed limb, where feeling is absent, and function is severely compromised. Not only can silenced places create local dysfunctions, they can also stir up noise elsewhere.

Sensory 'blind spots'

To be fully functional and at home in all of the body, awakening and integrating silenced places is vital. Be mindful, though; somatic awakening creates the perception of new, forgotten, suppressed, or avoided feelings, which can be elating, confusing, and frightening all at once. While it is still an undertaking worth engaging in, it is important to be mindful of the emotional soundscape that accompanies the process. Awakening numb areas without allowing time to make sense and integrate the emerging feelings might create a dissonant cacophony instead of our aim, a resonant somato-emotional symphony.

Awakening is vital for functionality and feeling at home in all of the body

Allow time for integrating new feelings

Gradual (Re)Integration of Feelings

It is much easier to integrate newly experienced feelings in small doses than in volumes. Think of a well-rehearsed group of musicians and then imagine a novice guitarist joining in. New to the group, the novice's skills are not yet fully refined. The music might be somewhat out of tune for a while. However, in time he will master his instrument and find his rhythm within the band. The group, with the guitarist well integrated, is primed to receive their next new musician with ease.

Now imagine you are playing or singing a song you really like. In an instant, ten people accompany you, each of them thrumming, strumming, and tooting on different instruments for the first time. They follow you around for the rest of the day. Sensory overload can feel like this!

Translating this illustration into a kinaesthetically-oriented movement practice means being patient and prioritising interoceptive clarity over simply feeling more.

When good things feel wrong, stay kind and curious

Kindness and Curiosity

Invaluable as somatic awakening and the expansion of sensory scope are, let's not idealise the journey. Experiencing new sensations and stirring up unexpected emotions can temporarily feel plain wrong; physically, mentally, and emotionally. For example, the absence of pain, discomfort, or even sensation in a silenced area may be interpreted as "everything is fine", whereas sensing something (anything) may translate into "something is seriously wrong". When a part of the body that was numb for some time starts to feel, it can be more disconcerting than elating at first. There is only one way to find out what the feeling means: stay with it, using kindness and curiosity.

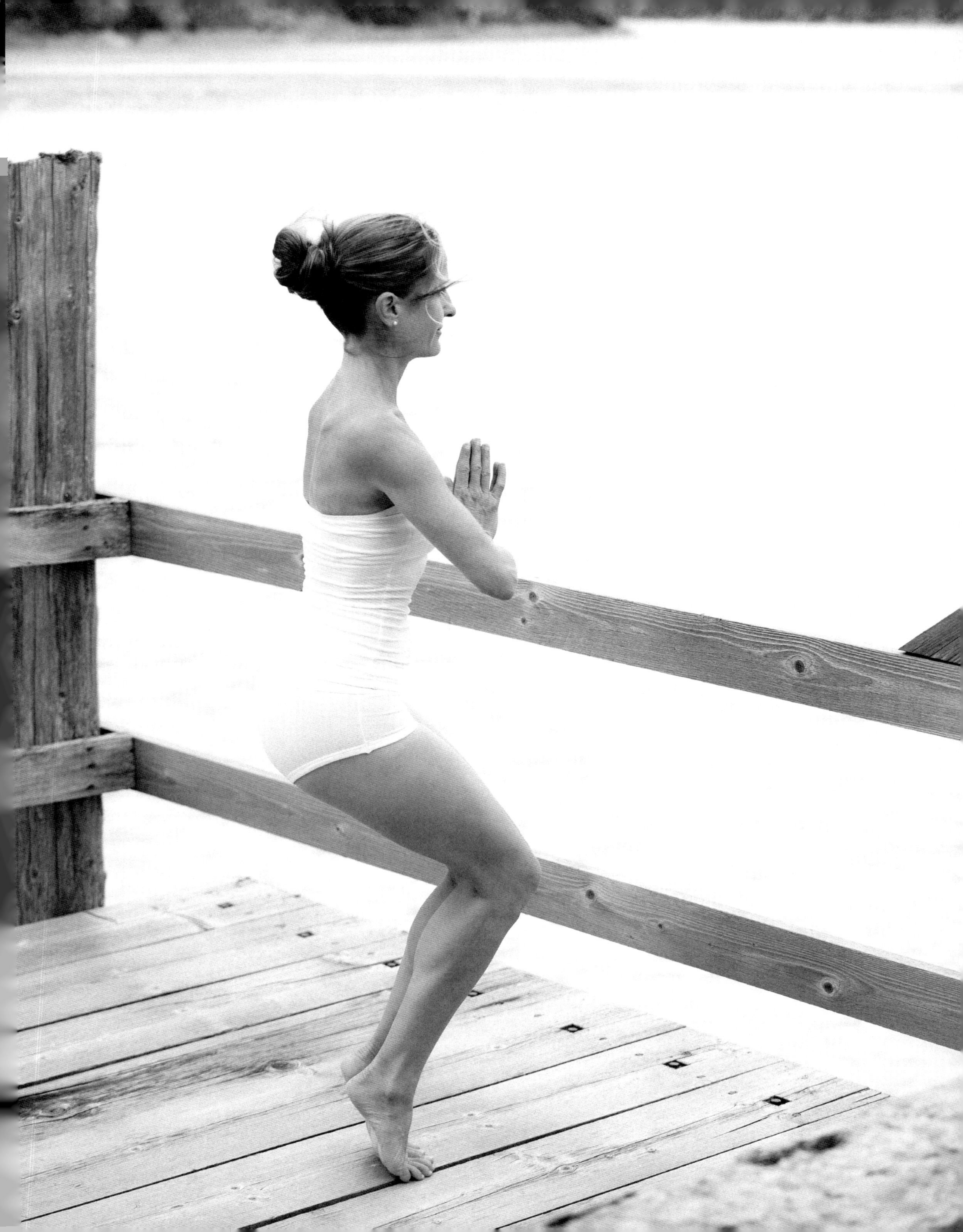

SLINGS IN MOTION

Kinaesthetic intelligence is not a ready-made gift; it is a skill that needs to be practised to be matured.

To train kinaesthetic intelligence, we employ both a proprioceptively-oriented and interoceptively-oriented methodology. By doing so, we sharpen intention and bring focus to kinaesthesia while engaging the other ten Fascial Movement Qualities.

Proprioceptively-Oriented Training

Focussing on body alignment, movement coordination, and rhythm

Proprioceptively-oriented training focusses on body alignment, movement coordination, and rhythm. As teachers, we can facilitate the process with literal verbal instructions and directing touch.

Description of actions:	Instructions describe what is done and how it is done.
Factual words:	The vocabulary and imagery relate directly to body alignment and movement coordination.
Sufficient detail:	Nuanced cueing that facilitates refined movement coordination by listening.
Technical descriptions:	Verbal cueing is instructive, including technical descriptions of movements (in a client-friendly language).
Directing touch:	Tactile instructions assist body alignment and movement coordination.
Direct translation:	Instructions can be directly translated into movement.
Question to ask:	How clearly do you align the body and coordinate movement?

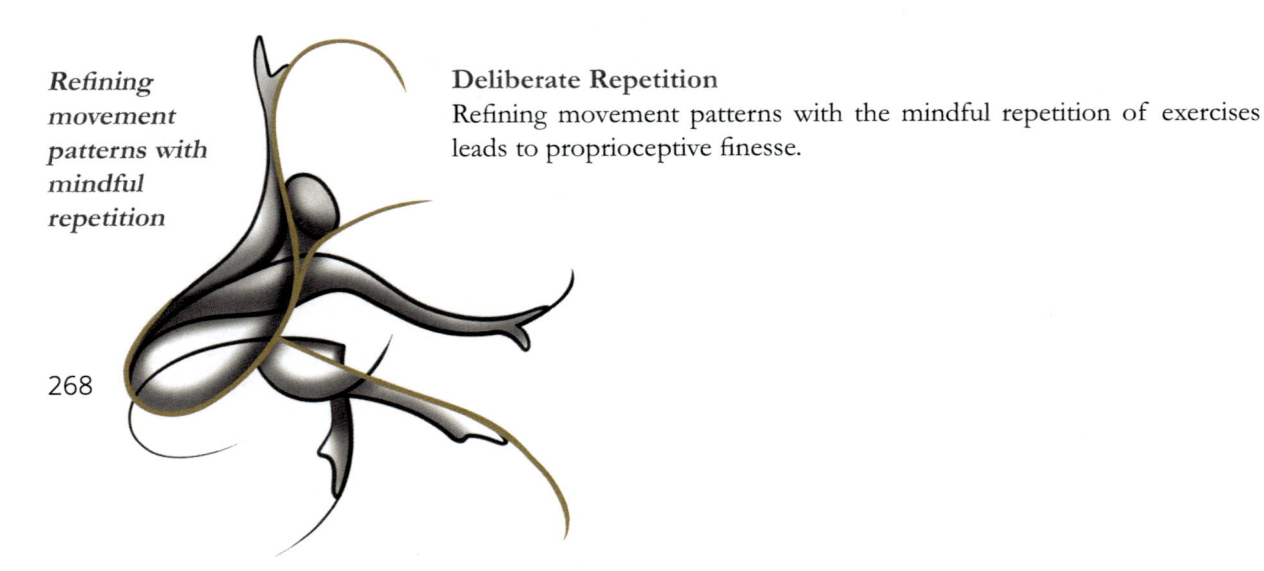

Refining movement patterns with mindful repetition

Deliberate Repetition
Refining movement patterns with the mindful repetition of exercises leads to proprioceptive finesse.

Interoceptively-Oriented Training

Interoceptively-oriented training focusses on the meaningful perception of physical sensations and affective states when moving. As teachers, we can facilitate the process with feeling words, imagery, and tactile guidance.

Focussing on the meaningful perception of physical sensations and emotional states when moving

Description of feelings:	Instructions are descriptive with respect to feelings while leaving room for personal interpretation.
Feeling words:	The vocabulary and imagery that relate to sensory and emotional experiences.
Space:	Quiet times for contemplation and internal answers are incorporated.
Guiding touch:	Tactile instructions convey the movement intention and quality.
Interpretive translation:	Instructions need to be interpreted to be translated into movement.
Question to ask:	How do you feel in your body, with your alignment and about the way you move?
Advanced question:	How do you feel about the way you feel?

Mindful Practice

Refining body awareness through mindfully executed movement and the embodiment of emotions leads to interoceptive clarity

Refining the quality of body and emotional awareness through mindful movement

PART 2

LANGUAGE MATTERS

Because language matters, we are mindful about the words we use when teaching Slings.

Sound of Words

Words have a certain feel

Consciously or not, we often associate words with certain feelings, which influences the exercise execution. There is no right or wrong wording, yet it is worthwhile to reflect on whether your vocabulary not only describes but also facilitates the desired movement quality on a sensory level. Compare the sounds of the following examples:

- Flex the spine versus curl the spine
- Twist the spine versus spiral the spine
- Pull the shoulder blades down the back versus glide the shoulder blades away from the ears
- Flex at the hips versus fold at the hips
- Contract or crunch the abdominals versus engage the abdominals
- Squeeze the pelvic floor versus draw up from the pelvic floor
- Hold the Plank or Body Tension versus maintain the Front Support

Active and Passive Wording

In Slings, the difference between active and passive wording is tremendously relevant to convey properties of fascial qualities correctly. Here a few examples:

Active: In a muscle-focussed knee flexion exercise, the wording is active, because muscles contract actively.
Cueing example: Bend the knee to a 90° angle.

Passive: In a fascia-focussed elastic movement against gravity, the wording is passive because fascia tensions and recoils on its own accord.
Cuing example: Let the knees bend and extend rhythmically.

Active: In a tensegral side bend that aims for active myofascial lengthening, the wording is chosen accordingly.
Cueing example: Lengthen the side of the body by pressing one foot down onto the floor and lifting the ear up toward the ceiling.

Passive: In a melting side bend, the wording is passive.
Cueing example: Let the side of your body melt like butter in the sun.

Active wording for actions to be done
Passive wording for actions that take place

INTROSPECTION, ABSORPTION, AND INVIGORATION

After elaborately discussing how we move and what motivates movement, let's allocate this space to the value of introspection, integration through stillness, and quiet invigoration of the senses.

Pause to Reflect

Pause the talking

During the practice, it is valuable to create space for the quiet reflection of internal states, may they be physical, mental, or emotional. When we 'pause to reflect', we don't pause the movement, we pause the talking, both externally and internally.

Absorption and invigoration

After the active part of the lesson, it is tremendously valuable to create space for integration through stillness. In Slings we call this time 'Absorption' because we let the body absorb and integrate structural changes and newly experienced feelings. Absorption is complemented by 'Invigoration', easeful movements that wake up the senses.

There is no strict formula for absorption and invigoration, though there is a format that has proven beneficial.

Richard and Anita with a group of body-minded movers in a state of absorption in Malaga in Spain

PART 2

Absorption

Absorption is a restful, integrative state of presence in which the body is still.

Restful state of presence in which the body is still

Supine: The body is at ease in a comfortable position laying on the back.

Restful: As long as it is restful, lie flat on the back. Props may also be used, such as a bolster beneath the knees or a pillow under the head.

Eyes closed: To calm the senses and turn the awareness inward, the eyes are closed.

Present: Body and mind are in a state of relaxed presence.

Mind anchors: Most of us urbanised mortals need more than a whispered suggestion to not be carried away by the constant stream of our thoughts when in a state of non-doing. That's why 'mind anchors' can be beneficial. By gently engaging the mind, it is anchored in the present, not rummaging in the past, fabricating a future, or escaping to castles in the sky. Mind anchors should have a somatic quality, which can either be consciously facilitated or take place without active contribution. Following are a few mind anchors that we use in Slings lessons:

- A guided journey through the body
- Guided breath awareness
- A piece of emotive music with meaningful lyrics
- A piece of writing that deepens the overall experience

Quiet: Silence is an essential part of the process. Either for the whole duration of Absorption or a few slow breaths after the 'mind-anchoring'.

Short timeframe: The usual timeframe ranges from five to ten minutes.

Absorption

Invigoration

Invigoration comprises easeful movement in which the senses are gently energised.

Sprawling:	Waking up the body with freestyle movements of the hands, feet, arms, and legs followed by pleasurable undulating stretching from a supine position is a joyful way to transition from stillness to activity.
Eyes open:	To reconnect with the outer world, the eyes are opened before moving into a sitting position.
Rocking and rolling:	A refreshing way of changing from supine into sitting is by curling into a ball and rocking back and forth. The invigorating effect is even greater when rolling generously, like in the exercise Rolling Like a Ball.
Spiralling and curling:	There are several ways to complete the 'Invigoration'. In sitting, two to four spiralling motions of the spine are complemented by one uplifting spine extension and one comforting spine flexion. The sequence is concluded in a centred upper body alignment.

Easeful movement in which the senses are gently energised

Invigoration:
Long Stretch   Rest Position   Rocking into Rolling   Rolling Like a Ball into Sitting

Spiralling Twist

Arch   Curl

Practising Quiet

Quiet is medicine for the mind

While some teachers and movers treasure the quiet times during the practice, others feel uneasy with the silence in the room. If unease is present, there is even more reason to facilitate undistracted introspection and integration through stillness. In a world where artificial noise engulfs us, and distraction is just the distance of your mobile phone away, quiet is medicine for the mind.

Teacher's tip: If silence or the discomfort of your participants when you are silent makes you restless and tempted to fill the space with words, commit to not speaking for ten slow counts. Consider it part of your practice to become comfortable with your own discomfort.

Mover's tip: If you like to do your personal practice with music playing or something else that entertains you, include five minutes of quiet time regularly. Commit in advance, otherwise the 'right' day for this new exercise might never come.

KINAESTHETIC INTELLIGENCE IMPROVES – EVERYTHING!

Kinaesthetic intelligence is trained in concert with all other Fascial Movement Qualities, which benefits every single one of them. In turn, the enhanced functionality of the other Fascial Movement Qualities strengthens our kinaesthetic intelligence. A cycle that is 360° life-quality enhancing!

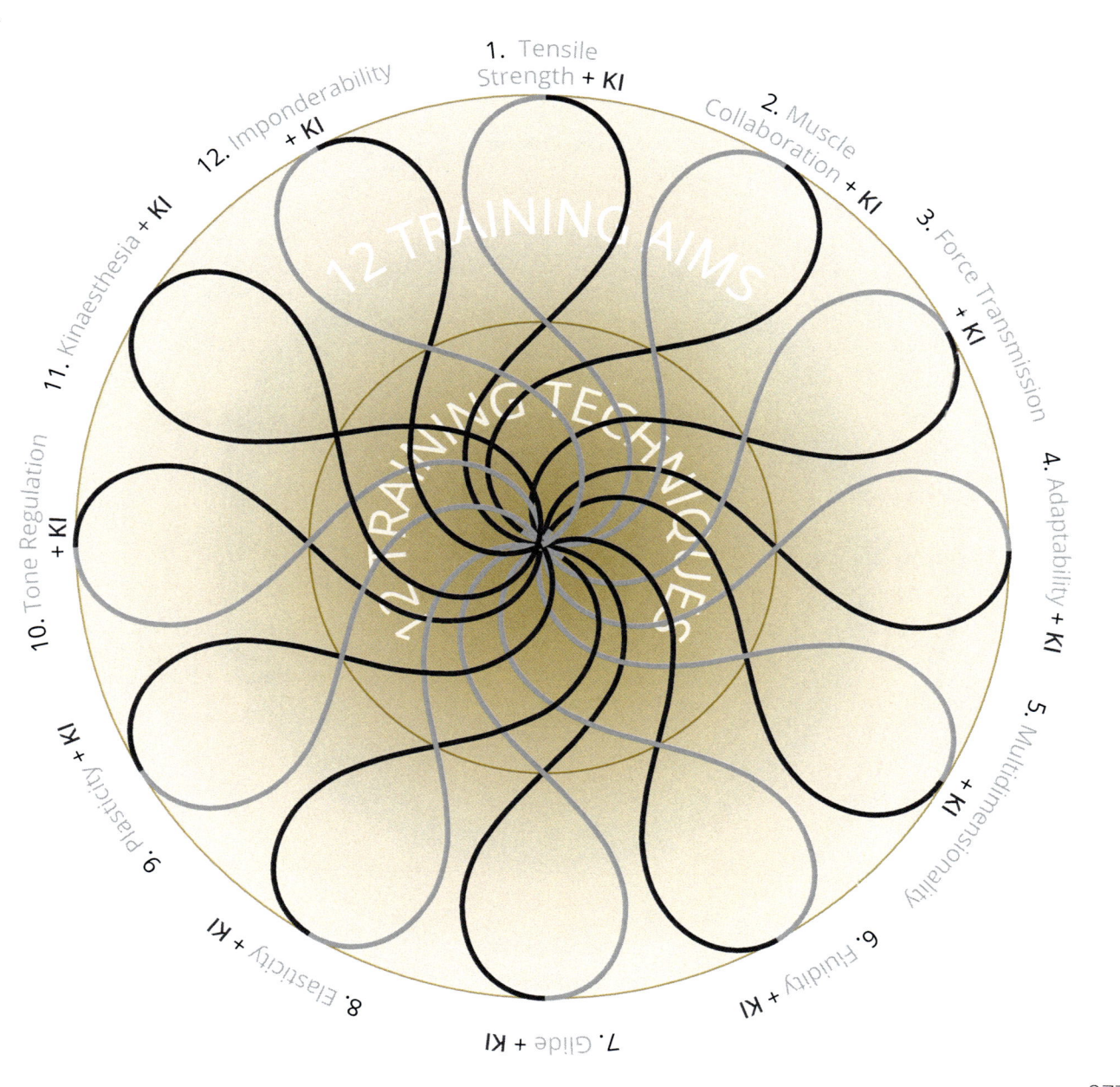

12 TRAINING AIMS / **12 TRAINING TECHNIQUES**

1. Tensile Strength + KI
2. Muscle Collaboration + KI
3. Force Transmission + KI
4. Adaptability + KI
5. Multidimensionality + KI
6. Fluidity + KI
7. Glide + KI
8. Elasticity + KI
9. Plasticity + KI
10. Tone Regulation + KI
11. Kinaesthesia + KI
12. Imponderability + KI

PART 2

Kinaesthetic Intelligence

Low Kneeling   Child's Pose

Sternum Massage

Prone Triple Extension

Small Wave

Transition: Front Support Inverted V

Forward Fold Leg Stretch   Rolling Up

Dynamic Double Knee Bend & Arm Pendulum

PART 2

Second Position Plié   Dynamic Plié

Dynamic Plié into Kite

Kite   Tree Pose

Transition: Kite into Second Position   *Side change*

Wide V Stance: Easeful Forward Fold   Deep Squat

Forward Fold Leg Stretch   Rolling Up   Standing Arch

ONTO THE MAT AND INTO THE BODY

Story and Practice: Kinaesthesia in Motion

SLINGS MYOFASCIAL TRAINING TECHNIQUES & AIMS

The following Slings Myofascial Training Techniques and Training Aims shown in gold, bold font are directly related to kinaesthesia and kinaesthetic intelligence.

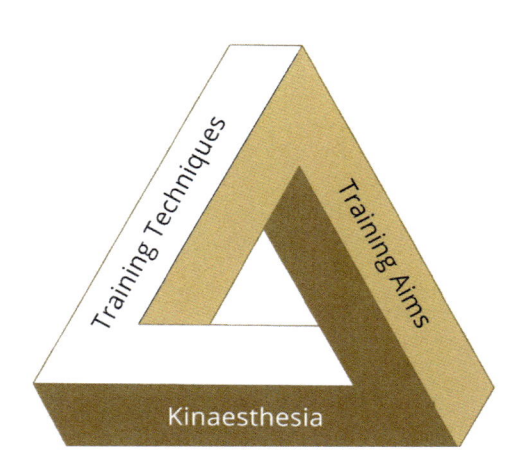

12 Myofascial Training Techniques

Techniques that optimise kinaesthesia:

1. **Stabilising**
2. **Toning**
3. **Pushing, Pulling, Counter-traction**
4. **Active Lengthening and Expansion**
5. **Spiralling**
6. **Hydrating**
7. **Gliding**
8. **Domino Motion**
9. **Bouncing and Swing**
10. **Massaging**
11. **Active Ease**
12. **Melting and Invigorating**

12 Training Aims

Twelve top achievements of kinaesthetic intelligence:

1. **Dynamic Stability**
2. **Multidimensional Strength**
3. **Limberness**
4. **Movement Rhythm**
5. **Elasticity**
6. **Tissue Nourishment**
7. **Adaptability**
8. **Resilience**
9. **Somatic Trust**
10. **Movement Courage**
11. **Mañana Competence**
12. **Kinaesthetic Intelligence**

IN A NUTSHELL
THINGS WORTH KNOWING ABOUT FASCIAL KINAESTHESIA AND MOVEMENT

4 Things About Feeling

Feeling:	Feeling encompasses sensation and emotion.
Sensation:	A sensation is the perception of something physical.
Emotion:	An emotion is the affective response to a sensation.
Correlation:	Sensations trigger emotions, and emotions change the quality of sensations.

4 Things About Kinaesthesia

Kinaesthesia:	Kinaesthesia means movement sense; it encompasses proprioception and interoception.
Sensory system:	Fascia can be viewed as the body's most influential perceptual system and the organ of kinaesthesia.
Proprioception:	The proprioceptive sense enables unconscious and conscious coordination of body alignment and movement.
Interoception:	The interoceptive sense facilitates the unconscious and conscious ability to feel the quality of physical sensations and their effect on emotional states, to adapt in a wellbeing-oriented manner.

4 Things About Kinaesthetic Intelligence

Kinaesthetic intelligence:	Kinaesthetic intelligence is the synergy of well-orchestrated movement, a vivid whole-body image, and embodied emotions.
Proprioceptive finesse:	Proprioceptive finesse is the ability to coordinate movement in a smooth and well-timed manner.
Interoceptive clarity:	Interoceptive clarity is the ability to sense the body with awareness to emotional states, and move in a wellbeing-oriented manner.
Dynamic balance:	Dynamic balance and interplay between proprioceptive finesse and interoceptive clarity characterise a person's kinaesthetic intelligence.

4 Things about Feeling Awareness

Informed choice-making: Understanding and embodying the different qualities of sensations and emotions enables choice-making in the body's best interest.

Noisy places: Noisy places are areas where intense feelings are experienced; for somatic ease, soothing and expanded awareness is required.

Quiet places: Quiet places are areas of somatic ease; appreciation strengthens inner resources and trust in the body.

Silenced places: Silenced places are sensory 'blind spots'; awakening and integrating these numb areas is vital for functionality and feeling at home in all of the body.

8 Slings Training Principles

Proprioceptively-oriented: Proprioceptively-oriented training focusses on body alignment, movement coordination, and rhythm.

Mindful repetition: Refining movement patterns with mindful repetition leads to proprioceptive finesse.

Interoceptively-oriented: Interoceptively-oriented training focusses on the meaningful perception of physical sensations and affective states.

Mindful movement: Refining the quality of body and emotional awareness through mindful movement leads to interoceptive clarity.

Language matters: Because language matters, the words to describe movements are chosen mindfully.

Introspection in motion: During the practice, space for silent introspection is deliberately included while the movement continues.

Absorption, invigoration: At the end of the practice, a time for absorption and integration in stillness is given. It is complemented with easeful movement in which the senses are gently invigorated.

Fascial Movement Qualities: Kinaesthetic intelligence is trained in concert with the other Fascial Movement Qualities.

3 BENEFITS OF FASCIAL KINAESTHESIA

1. Postural ease and well-orchestrated movement
2. Embodiment of emotions and wellbeing-oriented behaviour adaptations
3. Being and feeling vibrantly alive and whole

PART 2

The **imponderability** of your fascia is your space for wonder and discovery.

12. IMPONDERABILITY

"I would rather live in a world where my life is surrounded by mystery than live in a world so small that my mind could comprehend it."
Harry Emerson Fosdick

Imponderability refers to what is currently beyond our comprehension, even our imagination. The imponderable could be wholeness because wholeness can't be understood by the sum of its parts, as measurable and measured as the parts might be. It could also be the mystery that we start to see because of the knowledge or inner wisdom we acquire. Either way, the imponderable is the space for new explorations, scientific discoveries, and somatic insights. It is the summary of our experiences that, in unexplainable ways, can change the way we think, move, and feel.

ONTO THE MAT AND INTO THE BODY

Story and Practice: Imponderability in Motion

3 BENEFITS OF IMPONDERABILITY

1. Curiosity

2. Sense of wonder

3. Glad acceptance that some things are greater than what the mind can comprehend

12 SLINGS MYOFASCIAL TRAINING TECHNIQUES

Derived from the twelve Fascial Movement Qualities are the twelve Myofascial Training Techniques:

1. Stabilising
2. Toning
3. Pushing, Pulling, Counter-traction
4. Active Lengthening and Expansion
5. Spiralling
6. Hydrating
7. Gliding
8. Domino Motion
9. Bouncing and Swinging
10. Massaging
11. Active Ease
12. Melting and Invigorating

Application of the twelve Training Techniques facilitates the achievement of the twelve Training Aims.

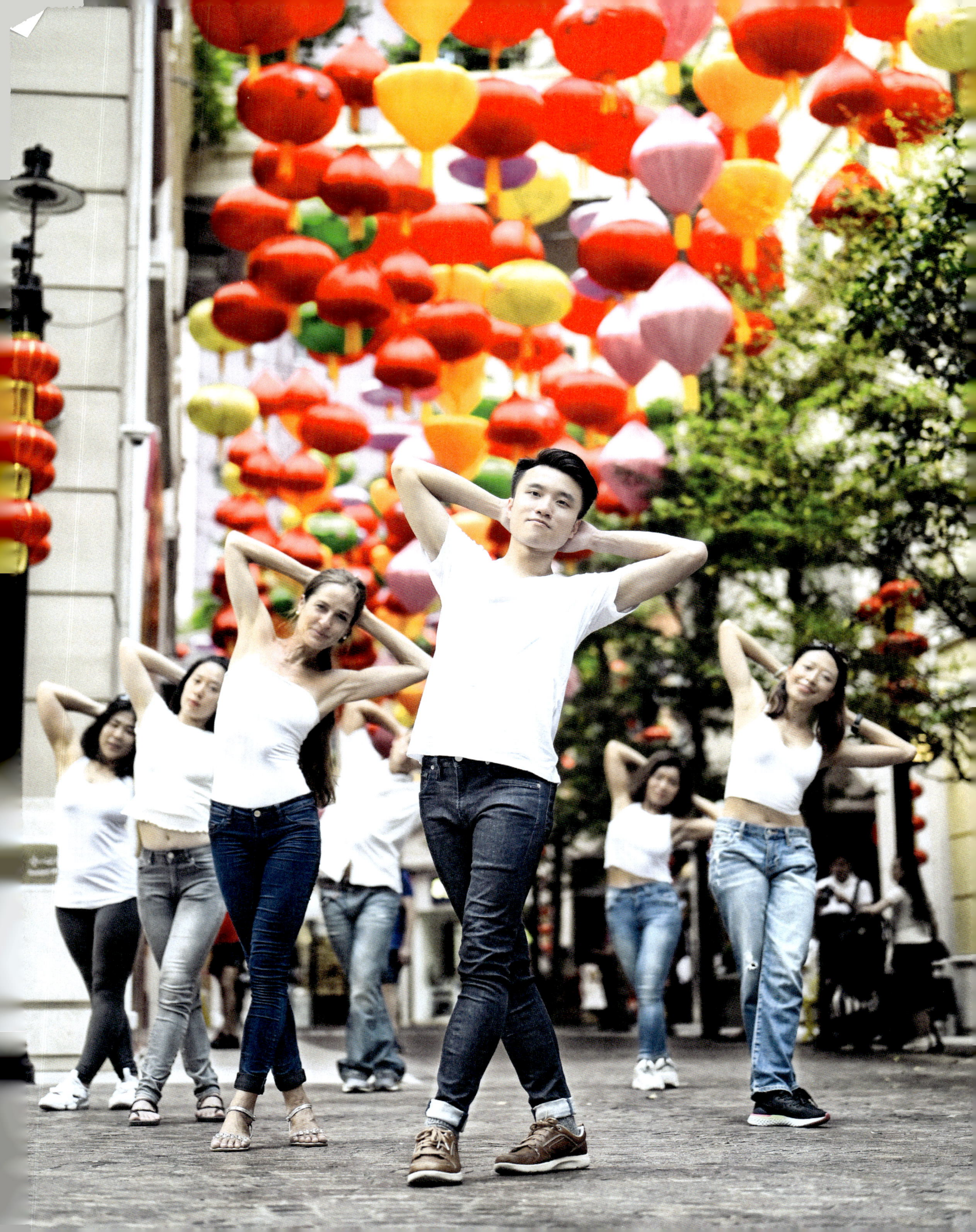

12 Fascial Movement Qualities and 12 Training Aims

Regarding the layout for the Training Techniques, we begin with a clear and informative description. Then, for each Training Technique, the Fascial Movement Qualities that serve as a more or less direct source of information are indicated. Also marked are the Training Aims facilitated by the application of a technique. The selections in the listings are not set in stone; instead, they serve as a flexible reference. After all, everything is connected. Definitions change with different viewpoints and contexts.

Finally, the Fascial Movement Qualities and Training Aims written in bold, gold letters are in focus, or most directly related to the discussed Training Technique. Regular black font indicates a contribution or positive side effect. Qualities and techniques written in grey colour are not in the immediate focus.

1. STABILISING

Focusses on engaging core stabilising muscles in a differentiated manner

As a technique, 'stabilising' refers to the deliberate strengthening of deep muscles that support joint alignment locally, in stillness and motion. It is applied to exercises for the whole body. That said, the central core is frequently in focus, therefore the lumbar pelvic area along with the bodywide myofascial core – the Deep Front Line.

Training dynamic stabilisation is an art and a skill that cannot be explained in a couple of sentences. Still, here are some pointers for you to consider when applying 'stabilisation' as a technique.

- When aiming for core activation, muscle engagement is minimal. The mantra is: "As much as necessary, as little as possible".

- Well-functioning muscles are not only strong, they are also sufficiently long and able to relax. These are qualities to be considered in practice.

- To minimise global muscle activation, the load placed on the body is fairly low in deep core stabilisation exercises.

- Regarding the trunk, the centred pelvic and spinal alignment is maintained. This means that the pelvis is stabilised in what is typically called a 'neutral' position – neither actively tilted anteriorly or posteriorly, while the natural curves of the spine are sustained throughout an exercise.

Training Techniques

- For the lower and upper extremities, balanced bone alignment is sustained statically and in motion.

- For bodywide dynamic stabilisation, we focus on optimal joint alignment in which muscle tone and fascial tension around a joint are balanced.

Primary Information Source
For additional information, refer to chapters one and two of the Fascial Movement Qualities: Tensile Strength and Muscle Collaboration.

12 Fascial Movement Qualities
Information sources:

1. Tensile Strength
2. **Muscle Collaboration**
3. Force Transmission
4. Adaptability
5. Multidimensionality
6. Fluidity
7. Glide
8. Elasticity
9. Plasticity
10. Tone Regulation
11. **Kinaesthesia**
12. Imponderability

12 Training Aims
'Stabilising' facilitates:

1. **Dynamic Stability**
2. Multidimensional Strength
3. Limberness
4. Movement Rhythm
5. Elasticity
6. Tissue Nourishment
7. Adaptability
8. Resilience
9. **Somatic Trust**
10. Movement Courage
11. **Mañana Competence**
12. **Kinaesthetic Intelligence**

4-Point Leg Press

PART 2

2. TONING

Focusses on strengthening movement-oriented muscles in a diverse manner

As a technique, 'toning' refers to deliberate strengthening of global, movement-oriented muscles.

Toning exercises include isometric, eccentric, and concentric muscle actions. They are executed in all body positions, in all planes of movement, with different intensities, and in various rhythms. Muscle toning exercises are as diverse as fascia-focussed movements and just as important.

Primary Information Source
For additional information, refer to chapter two of the Fascial Movement Qualities: Muscle Collaboration.

12 Fascial Movement Qualities
Information sources:

1. Tensile Strength
2. **Muscle Collaboration**
3. Force Transmission
4. Adaptability
5. Multidimensionality
6. Fluidity
7. Glide
8. Elasticity
9. Plasticity
10. Tone Regulation
11. **Kinaesthesia**
12. Imponderability

12 Training Aims
'Toning' facilitates:

1. Dynamic Stability
2. **Multidimensional Strength**
3. Limberness
4. Movement Rhythm
5. Elasticity
6. Tissue Nourishment
7. Adaptability
8. Resilience
9. **Somatic Trust**
10. **Movement Courage**
11. Mañana Competence
12. **Kinaesthetic Intelligence**

Cleopatra   Kneeling Side Bend   Kneeling Side Support with Leg Lift & Spiralling Leg Kick Front

3. PUSHING, PULLING, COUNTERTRACTION

As a technique, 'pushing, pulling, counter-traction' refers to muscle actions that engage specific myofascial slings deliberately or change the way they work. Oftentimes these actions are isometric or nearly isometric, though the technique is also applied to muscles in motion, most commonly in eccentric movements when muscles actively lengthen.

'Pushing and pulling' are usually done with the hands or the feet against the floor, or an immovable surface. Outwardly the body position doesn't change; variations in muscle tone and fascial tension are purely internal.

In 'counter-traction' a myofascial sling is actively lengthened in opposing directions. The technique can be applied to one joint, a body part, or whole myofascial meridians.

Primary Information Source
For additional information, refer to chapter three of the Fascial Movement Qualities: Force Transmission.

Focusses on internal activation or balanced active lengthening of myofascial slings

90/90 Side Bend

12 Fascial Movement Qualities
Information sources:

1. Tensile Strength
2. **Muscle Collaboration**
3. **Force Transmission**
4. Adaptability
5. Multidimensionality
6. Fluidity
7. Glide
8. Elasticity
9. Plasticity
10. Tone Regulation
11. **Kinaesthesia**
12. Imponderability

12 Training Aims
'Pushing, pulling, counter-traction' facilitates:

1. Dynamic Stability
2. **Multidimensional Strength**
3. **Limberness**
4. Movement Rhythm
5. Elasticity
6. Tissue Nourishment
7. Adaptability
8. Resilience
9. Somatic Trust
10. Movement Courage
11. Mañana Competence
12. **Kinaesthetic Intelligence**

PART 2

4. ACTIVE LENGTHENING AND EXPANSION

Focusses on structural decompression and inner spaciousness

As a technique, 'active lengthening and expansion' refers to the deliberate elongation of the trunk and limbs, as well as an increase in volume in the upper body. The resulting inner spaciousness decompresses the body.

'Active lengthening' frequently focusses on spinal elongation to increase the space between the pelvis and head. It is also used for the extremities, from the shoulder girdle to the fingertips, and from the pelvis to the toes.

Favourite areas for 'expansion' are the back of the pelvis, the lower back, and the entire ribcage.

A deep inhalation is commonly emphasised to enhance both active lengthening and expansion.

Primary Information Source
For additional information, refer to chapter one of the Fascial Movement Qualities: Tensile Strength.

12 Fascial Movement Qualities
Information sources:

1. **Tensile Strength**
2. Muscle Collaboration
3. Force Transmission
4. Adaptability
5. **Multidimensionality**
6. Fluidity
7. **Glide**
8. Elasticity
9. Plasticity
10. **Tone Regulation**
11. **Kinaesthesia**
12. Imponderability

12 Training Aims
'Active lengthening and expansion' facilitates:

1. Dynamic Stability
2. Multidimensional Strength
3. **Limberness**
4. Movement Rhythm
5. Elasticity
6. Tissue Nourishment
7. Adaptability
8. **Resilience**
9. Somatic Trust
10. Movement Courage
11. Mañana Competence
12. **Kinaesthetic Intelligence**

Kneeling Triple Extension

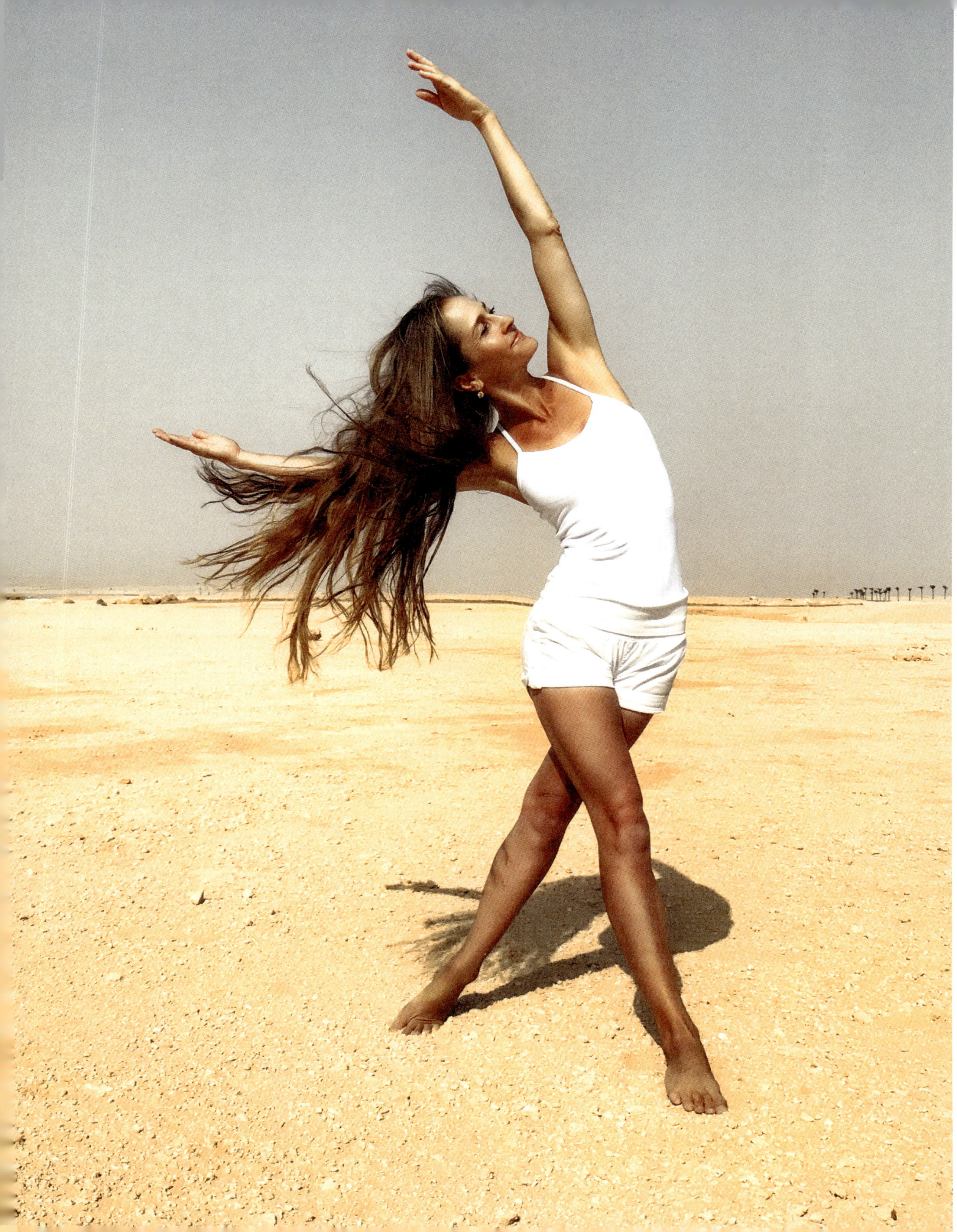

5. SPIRALLING

Focusses on multidimensional movements

As a technique, 'spiralling' refers to the deliberate execution and sensing of three-dimensional, rotational movements.

When 'spiralling' the spine, it lengthens and, at the same time, rotates around its own axis. Usually, this is done with an expansive inhalation. With the exhalation, the spine turns back to its original position. Spiralling of the elongated spine often, though not always, utilises a degree of fascial elasticity.

The ribcage is also a favourite place of spiralling motions. Ribcage spiralling encompasses two components: one is slight rotational joint movement, and the other is a sense of allowing the ribs to turn within their fascial envelopes. Spiralling of the ribs is often emphasised in multidimensional movements where the spine is first side bent and then rotated. With practice, ribcage spiralling can also be perceived in the previously described spiralling of the spine.

The concept of 'spiralling' is also applied to the shoulders and hips. In these joints, there is a sense of increased spaciousness during the rotational phase of the motion, complemented with the feeling of a deepening connection in the de-rotation phase.

Primary Information Sources
For additional information, refer to chapters one and five of the Fascial Movement Qualities: Tensile Strength and Multidimensionality, respectively.

12 Fascial Movement Qualities
Information sources:

1. **Tensile Strength**
2. Muscle Collaboration
3. Force Transmission
4. **Adaptability**
5. **Multidimensionality**
6. Fluidity
7. **Glide**
8. Elasticity
9. Plasticity
10. **Tone Regulation**
11. **Kinaesthesia**
12. Imponderability

12 Training Aims
'Spiralling' facilitates:

1. **Dynamic Stability**
2. **Multidimensional Strength**
3. **Limberness**
4. Movement Rhythm
5. Elasticity
6. **Tissue Nourishment**
7. **Adaptability**
8. **Resilience**
9. **Somatic Trust**
10. **Movement Courage**
11. **Mañana Competence**
12. **Kinaesthetic Intelligence**

Training Techniques

Dynamic Hip Release & Turning In and Out

Dynamic Hip Release & Spiralling

Triangle Side Stretch & Spiralling

6. HYDRATING

Focusses on fluid flow and tissue nourishment

Small Wave & Kneeling Upward Stretch into Child's Pose

As a technique, 'hydrating' refers to the deliberate promotion of fluid flow in fascia.

In exercises without props, movement interplays that actively lengthen and soften muscles and fascia are most commonly used to cleanse and rehydrate the tissues, respectively.

Self-massage exercises utilise alternated pressure and release for the same purpose.

The idea is based on the 'sponge principle'. During the active lengthening, or pressure phase, fluid, and with it waste products, is squeezed out of the fascia. By subsequently softening the myofascial structures, fresh fluid, and with it, nutrients, is drawn in and absorbed. This nourishes tissues and enhances fluid flow immediately and in the hours to come.

Most commonly, 'hydrating' is applied to active lengthening and self-massage repertoire. However, the concept can also be applied to toning exercises. For movement-based hydration, the following guidelines apply:

- Active lengthening is followed by softening into the opposite movement direction.
- Toning is followed by softening into the same movement direction

Primary Information Sources

For additional information, refer to chapter six of the Fascial Movement Qualities: Fluidity.

12 Fascial Movement Qualities
Information sources:

1. Tensile Strength
2. Muscle Collaboration
3. Force Transmission
4. Adaptability
5. Multidimensionality
6. **Fluidity**
7. Glide
8. Elasticity
9. Plasticity
10. Tone Regulation
11. **Kinaesthesia**
12. Imponderability

12 Training Aims
'Hydrating' supports:

1. Dynamic Stability
2. **Multidimensional Strength**
3. **Limberness**
4. Movement Rhythm
5. Elasticity
6. **Tissue Nourishment**
7. Adaptability
8. Resilience
9. **Somatic Trust**
10. Movement Courage
11. **Mañana Competence**
12. **Kinaesthetic Intelligence**

Training Techniques

12 Fascial Movement Qualities
Information sources:

1. Tensile Strength
2. Muscle Collaboration
3. Force Transmission
4. Adaptability
5. **Multidimensionality**
6. **Fluidity**
7. **Glide**
8. Elasticity
9. Plasticity
10. **Tone Regulation**
11. **Kinaesthesia**
12. Imponderability

12 Training Aims
'Domino motion' enhances:

1. Dynamic Stability
2. Multidimensional Strength
3. **Limberness**
4. **Movement Rhythm**
5. Elasticity
6. **Tissue Nourishment**
7. **Adaptability**
8. Resilience
9. Somatic Trust
10. Movement Courage
11. **Mañana Competence**
12. **Kinaesthetic Intelligence**

Dynamic Mermaid

9. BOUNCING AND SWINGING

Focusses on rhythmical, elastic motion

As a technique, 'bouncing and swinging' refers to rhythmical movements in which fascia elastically lengthens and recoils. Both swinging and bouncing motions are commonly performed dynamically against gravity, though there are exceptions.

In bouncing exercises that focus on or involve elasticity in the lower body, there is a sense of 'letting the joints fold', which results in the elastic lengthening of associated fascia. When the tensioned fascial structures elastically recoil, the body is propelled back towards its starting position. By letting the arms swing dynamically and with minimal muscular effort, elasticity in the shoulders is enhanced, and momentum is gained for other elastically moving body parts. As for elasticity in the back, swinging exercises in all planes of movement are progressively introduced to the body.

Although the goal is to use minimal muscular effort when 'bouncing and swinging', we always need some – and occasionally some more – muscular energy to utilise fascial elasticity on the mat. Especially when the range of movement is large, a body part is heavy, or when the action is not supported by gravity, we add a pinch of extra muscular energy to move safely and efficiently.

The dynamic rhythm of the exercises matches the body part and the tensile strength of the mover's fascia. When the motion feels springy and buoyant, the rhythm is optimal.

Primary Information Source
For additional information, refer to chapter eight of the Fascial Movement Qualities: Elasticity.

Training Techniques

12 Fascial Movement Qualities
Information sources:

1. **Tensile Strength**
2. Muscle Collaboration
3. Force Transmission
4. Adaptability
5. Multidimensionality
6. Fluidity
7. Glide
8. **Elasticity**
9. Plasticity
10. Tone Regulation
11. **Kinaesthesia**
12. Imponderability

12 Training Aims
'Elasticity' facilitates:

1. Dynamic Stability
2. Multidimensional Strength
3. **Limberness**
4. **Movement Rhythm**
5. **Elasticity**
6. Tissue Nourishment
7. **Adaptability**
8. **Resilience**
9. Somatic Trust
10. **Movement Courage**
10. Mañana Competence
11. **Kinaesthetic Intelligence**

Energy Swing

10. MASSAGING

Focusses on stimulating and hydrating fascia with self-massage

As a technique, 'massaging' refers to the use of props to stimulate and hydrate fascia with self-massage exercises.

The pressure can be feather-light to enliven the superficial fascia just beneath the skin and the skin itself. It can also be firm and deep to invigorate myofascial structures that are more profound (only in terms of location, not importance, though!) – to only do one or the other misses half of the story.

When aiming for stimulation, the massaged body part can be relaxed or engaged. In that sense, the muscles relaxed or firm, and the fascia softened or tensioned. The pace can range from a casual slow pace to speedy rolling.

If hydration is in focus, the treated part of the body is as relaxed as possible during the massage. In this way, the fascial tension and muscular tone are minimal. It allows for a deeper pressure that reaches more tissues with the least amount of structural resistance. Relaxation is not possible in all exercises though. In such cases, it is beneficial to adopt a relaxed counter-pose in which the tissues are softened, subsequent to the massage. The rolling pace ranges from slow to very slow. Pressure can also be sustained. Ten to twenty seconds is typically sufficient.

In all self-massage exercises, the breath must flow; holding the breath is counterproductive to our aim, which is vitalising myofascial structures and enhancing fluid flow. During the massage, the mover ideally feels that they are rolling towards wellbeing with a sense of pleasure; in contrast to getting through the exercise by grinding the teeth and triggering a spike in sympathetic nervous system activation.

Note that if the props used are textured, they might leave marks on the skin for a short while; that is okay. Bruising on the other hand, is not okay. We don't aim to damage the body, but support healing and health. Should there be bruising, it is a sign that the pressure was too intense, the prop too hard, or the duration too long. Lighten the pressure, soften the prop, and shorten the duration in the next session.

Training Techniques

Primary Information Sources

For additional information, refer to chapters six and eleven of the Fascial Movement Qualities: Fluidity and Kinaesthesia.

12 Fascial Movement Qualities

Information sources:

1. Tensile Strength
2. Muscle Collaboration
3. Force Transmission
4. Adaptability
5. Multidimensionality
6. **Fluidity**
7. **Glide**
8. Elasticity
9. **Plasticity**
10. Tone Regulation
11. **Kinaesthesia**
12. Imponderability

12 Training Aims

'Massaging' facilitates:

1. Dynamic Stability
2. Multidimensional Strength
3. **Limberness**
4. Movement Rhythm
5. Elasticity
6. **Tissue Nourishment**
7. **Adaptability**
8. Resilience
9. Somatic Trust
10. Movement Courage
11. Mañana Competence
12. **Kinaesthetic Intelligence**

Neck Massage with Nodding

11. ACTIVE EASE

Focusses on light activity in one body part to aid easeful change in another

As a technique, 'active ease' primarily refers to the relatively passive mobilisation of joints, or an easeful gradual lengthening or softening of myofascial structures. The mobilisation, lengthening, or softening in one part of the body is actively supported by another body part. For example, an arm is used to move a leg, or the leg muscles work as much as needed to enable the upper body to fully relax.

To decrease activity in the body part that is released, props such as a strap can be used to guide the movement. In some exercises, balls, domes, or yoga blocks can assist the opening of joints; therefore, myofascial changes in a non-strenuous manner, facilitate fluid flow and unload myofascial structures as well as organs.

'Active ease' can be applied to dynamically performed movements or sustained poses.

Primary Information Source
For additional information, refer to chapter nine of the Fascial Movement Qualities: Plasticity.

12 Fascial Movement Qualities
Information sources:

1. Tensile Strength
2. Muscle Collaboration
3. Force Transmission
4. Adaptability
5. Multidimensionality
6. Fluidity
7. Glide
8. Elasticity
9. Plasticity
10. **Tone Regulation**
11. **Kinaesthesia**
12. Imponderability

12 Training Aims
'Active ease' facilitates:

1. Dynamic Stability
2. Multidimensional Strength
3. **Limberness**
4. Movement Rhythm
5. Elasticity
6. Tissue Nourishment
7. Adaptability
8. Resilience
9. Somatic Trust
10. Movement Courage
11. **Mañana Competence**
12. **Kinaesthetic Intelligence**

Training Techniques

Dynamic Adductor Stretch

Easeful Adductor Stretch

Focusses on creating change through easeful motion or restful poses followed by light activity

12. MELTING & INVIGORATING

As a technique, 'melting and invigorating' refers to deliberately enhancing fascial permeability in an easeful manner in combination with stimulation of muscles and fascia to assure sufficient physical responsiveness and resilience.

Melting or the sense of melting are promoted with sustained, quiet poses that feel restful. Enjoyable sustained or slow-moving massages as well as pleasurable, glide-enhancing motions are incorporated.

In melting poses, muscles and fascia gradually lengthen in a relaxed manner. The average time frame of melting poses ranges from one to five minutes. Invigorating motions gently engage and soften the previously lengthened muscles and fascia in an easeful, dynamic manner.

Melting massage comprises easeful slow-rolling or a comfortable degree of sustained pressure. Subsequent to the latter, a gently invigorating movement is recommended.

Melting in motion often focusses on generous, glide-enhancing exercises that comprise spiralling, circling, and wave-like elements.

Primary Information Source
For additional information, refer to chapter nine of the Fascial Movement Qualities: Plasticity.

Training Techniques

12 Fascial Movement Qualities
Information sources:
1. Tensile Strength
2. Muscle Collaboration
3. Force Transmission
4. **Adaptability**
5. Multidimensionality
6. **Fluidity**
7. **Glide**
8. Elasticity
9. **Plasticity**
10. **Tone Regulation**
11. **Kinaesthesia**
12. Imponderability

12 Training Aims
'Melting and invigoration' facilitates:
1. Dynamic Stability
2. Multidimensional Strength
3. **Limberness**
4. Movement Rhythm
5. Elasticity
6. **Tissue Nourishment**
7. **Adaptability**
8. Resilience
9. Somatic Trust
10. Movement Courage
11. **Mañana Competence**
12. **Kinaesthetic Intelligence**

Melting Deer Pose

Mindful transition into Z Sit

Sitting Hip Opener

ALL IN ONE

Every exercise utilises more than one technique. It is up to you to decide which technique or techniques you want to focus on in each movement. Define your priorities according to the current lesson topic or selected training aims, as well as the degree of exercise embodiment. For example, if the lesson topic is 'enhanced glide', select a good number of exercises in which the 'gliding' technique is applied. Prioritise cueing 'gliding' and its purpose over other techniques whenever applicable.

For new exercises, it is recommended to focus on one or two techniques only. This makes comprehension and integration more accessible, sharpening the movement intention and deep effects. The more embodied an exercise is, the more the awareness can be brought to different techniques at different movement phases or in different body parts.

Training Techniques

Karin, Adrian, and Muriel embodying the Slings Myofascial Training Techniques with ease on a salt lake in Western Australia

12 SLINGS TRAINING AIMS

While every exercise has its specific goals, the practice as a whole has twelve defined Training Aims:

1. Dynamic Stability
2. Multidimensional Strength
3. Limberness
4. Movement Rhythm
5. Elasticity
6. Tissue Nourishment
7. Adaptability
8. Resilience
9. Somatic Trust
10. Movement Courage
11. Mañana Competence
12. Kinaesthetic Intelligence

1. DYNAMIC STABILITY

Dynamic stability is the process of maintaining structural integrity 24/7. In a dynamically stabilised body, posture is decompressed. There is a sense of ease when standing in line at a counter or even sitting in an office chair for hours. Movement can be performed with greater safety and efficiency, which promotes somatic trust and movement courage. Aside from the physical advantages, dynamic stability can foster a sense of inner strength and a feeling of "having it together".

In practice, dynamic stability allows the mover to explore healthy ranges of movement and multidimensionally, with confidence and resilience. Often this is where the challenge zone can become the play zone, and eventually even the comfort zone. Developing dynamic stability cannot be rushed. It is well worth the effort(s).

2. MULTIDIMENSIONAL STRENGTH

When the muscular system can handle strength-related tasks with efficiency, regardless of the body position or joint alignments, it is not only useful; it feels really good! Multidimensional strength enables you to function well, with a degree of ease when lifting a sleeping baby out of the cot, walking up the stairs or a mountain, or when carrying groceries. In one way or another, it makes every day a little easier; even a strenuous task, can become a little joyful.

Multidimensional strength is a prerequisite for many of the movements in the Slings repertoire, particularly those where rhythm and weight-bearing unite – regardless of body position or joint alignments. Additionally, when truly multidimensional and harmoniously balanced, strength lends to overall movement ease, instilling somatic trust on and off the mat.

3. LIMBERNESS

Limberness unites balanced joint mobility, muscular flexibility, and fascial adaptability within healthy parameters. Having a bodywide range of movement that matches everyday demands supports all-around functionality and the longevity of joints. It also enables you to move in response to your wants and needs, not according to your limitations. You can bend down to tie your shoelaces, reach up to the top shelf of a cupboard, or squat down to power up without thinking twice. A feel-good degree of limberness can also facilitate a sense of inner freedom.

This inner freedom that parallels limberness develops over time with consistent practise. It is important to remember that limberness and flexibility mean different things (Temple Dancers beware!). This integration of a healthy dose of balanced joint mobility, muscular flexibility, and fascial adaptability may be where your Temple Dancer embraces their inner Viking – and vice versa. Remember, with Slings, the magic is in the mix – the exercise selection and the sequencing. As limberness develops, that mix has more potential to bloom and grow, begetting more challenge and play along the way.

4. MOVEMENT RHYTHM

Appropriately timed movement is part of being well-coordinated with grace, the kind of grace that shows in the fluid way animals move, or skilled athletes and dancers perform. It enables you to orchestrate your activities in an energy-efficient manner that is coherent with the larger movement context. Well-timed movement rhythm also adds clarity and expressiveness to your body language. Besides, the speed or slowness in which you move changes the way you perceive yourself and the outside world.

If movement rhythm seems foreign or clumsy at first in practise, Slings provides many opportunities to engage your own inner and outer, beautifully varying song. From the flowing movement of a 'bone domino' to the more dynamic bounces and swings, the mat may provide an oasis where exploration of your inner grace, unique body language, and timing can find depth and confidence. Movement Rhythm is a worthwhile consideration for those of us whose minds are racing or hibernating in a state of inertia.

5. ELASTICITY

Fascial elasticity adds spring to rhythmical movements while conserving muscular energy. You feel the energetic difference in walking where one hour of stop-start marching through busy city streets can feel more tiresome than three hours of continuous wandering in nature. The elastic energy contribution of fascia also makes a perceptible contribution when running for gold or after the kids, leaping for joy or over hurdles, kicking a ball, swinging a tennis racket, golf club, or the arms, and of course any dance floor in the world.

The energy contribution of fascia is as perceptible in practice as it is in life. This training aim requires time to embrain and embody, as well as patience, to realise in movement. At first, movement sequences that indicate elastic qualities may not feel so elastic. Still, enjoy the journey to more buoyancy. In time, it not only adds power and ease to your activities and practice; it also feels really good physically and emotionally. Movement love is guaranteed.

6. TISSUE NOURISHMENT

When moving in diverse ways, fluid is circulated throughout the fascial system. The fluid flow continuously washes waste products out and brings fresh nutrients to the tissue. In turn, well-nourished fascia supports the health of the muscles and other organs embedded in it, as well as the vigour of cells.

Movement is as nutritious to our tissues, as sustenance is to our appetite. Our bodies may not generate hunger pangs for physical activity, yet they do communicate longing for motility in various ways. A pandiculating cat getting up from a mat demonstrates this inner longing beautifully. Tune in to your own inner cravings for tissue nourishment and weave fluidity enhancing exercises into your practice. The enhanced somatic juiciness is an energy booster that facilitates movement ease, radiant vitality, and the feeling of being at home in all of the body.

7. ADAPTABILITY

The unconscious or conscious ability to respond to changing circumstances increases physical agility and prevents injury. An adaptable body responds effectively when stumbling over a curb, missing a step on the staircase, twisting the ankle on the soccer field, or turning awkwardly when catching the falling soap in the shower. Fascial adaptability prepares for the unprepared. It makes you more resilient, which brings a sense of ease to body and mind.

The diversity of the Slings exercise repertoire, the contrasting sequences, and lessons that ebb and flow both lay the foundation for adaptability and develop it. It is wise to acknowledge the value of time with this aim. Many elements come together to create it. Just watch a toddler as they teeter and wobble along, then not too far down the road the same teenager is bending it like Beckham on the football field. That is maturing adaptability in motion! It also requires a lot of learning and practice along the way. Spiralling, gliding, melting, bouncing, and swinging allow us to meet the mosaic of challenges and joys that life throws our way. As our physical and emotional adaptability matures, we learn to take events in, respond, and act, attenuating to the needs of each situation. A practice that includes exercises that improve adaptability serves us very well in everyday life.

8. RESILIENCE

Resilience is the ability to bounce back from adversity. This capacity to rise to meet challenges with sufficient stamina and self-healing, as well as the possibility to psycho-emotionally cope with and recover from hardship, is not a trait that you have or don't have. It is an ability that involves actions, thoughts, and behaviours that can be developed and strengthened in everyone. By training the body's largest sensory system – fascia – we train our resilience.

For those who might believe they 'don't' have resilience, the mat can be an excellent place to experiment and push limits. A curious mind that questions interoceptive urges to come out of a challenging pose or avoid certain movements may wisely lean in and instead develop a sense of mastery over what once seemed impossible. Equally important is the mover who perhaps has always pushed through pain, even to injury, who learns over time to listen to and trust their own somatic wisdom.

Having resilience fosters somatic trust and movement courage, which are essential ingredients for sustained wellbeing. What having resilience doesn't mean is that experiencing pain is easy. Instead, it means that you can meet challenges well-resourced and recover within a healthy time frame, physically and emotionally.

9. SOMATIC TRUST

Somatic trust means feeling confident in your body's ability to reliably function and cope with challenges. Both of which are essential for maintaining a sense of inner strength, trust in your self-healing capacity, and for sustaining or regaining movement courage. Perpetuating a feeling of fragility can lead to overly protective behaviour, the fear to move freely, or move at all. Somatic trust unconsciously or consciously supports the willingness and ability to move in diverse ways, which strengthens resilience, and in turn, facilitates self-healing. It is a powerful sign of somatic trust when you stay active during or after physical challenges, move smartly when you think you can't or shouldn't, and try something new to overcome self-imposed limitations.

Somatic trust is a delightful quality to recognise within the moment it takes flight. Your inner witness suddenly notices that something once avoided is now accessible – this can be fun, and at times, even thrilling. At first, these moments may be inconsistent as the body tests the waters and develops confidence for various movements and situations. Allow your body to take the time it needs to earn its own trust and see where that takes you. Somatic trust is a profound training aim that may pleasantly surprise you on and off the mat.

10. MOVEMENT COURAGE

When voluntarily stepping out of your comfort zone, and even out of the play zone, to explore the challenge zone, you are exercising movement courage. It is the ability to engage in new or different exercises that make your heart beat a little faster, activities that you believe to be unavailable or no longer available, or movements you deliberately avoid. Movement courage is not adrenalin or attention-seeking movement bravado, though – quite the contrary. At the heart of movement courage is mindfulness and humility. It is the willingness to fail and then do it again – and again …

Physical and emotional resilience, as well as somatic trust, support movement courage. In combination, these qualities keep your activity spectrum broad and diverse, which is especially important as we age. They also offer the possibility to recognise and utilise unused movement potential that might come in handy when you least expect it. Movement courage can deliberately shake up inertia on the mat or excessive repetitiveness in everyday life. The exhilaration and sense of achievement that comes from mastering something new or different can invigorate your practice as much as the other waking hours of your day.

11. MAÑANA COMPETENCE

Mañana competence can be practised in movement or stillness. The inner quietude it fosters enables you to slow down and let the parasympathetic nervous system take over. This shift promotes rejuvenation, recovery, and the ability to be present and experience feelings vividly.

In contrast, rushing through practice, or life, with a haunting sense of tension, decreases enjoyment. In the long run, this is unhealthy, fogs the thinking, and causes a loss of inner and outer perspective. In addition to the immediate benefits of mañana competence, carrying the serenity and grounding of your movement practice into the future creates the ability to remain calm when life is turbulent.

Slings practice provides a variety of opportunities to experiment with and try different ways of strengthening your mañana competence. For some of us, on the mat may be the only place where we can begin to recognise this valuable sense of peace. From pausing in a moment of challenge and perhaps hanging in for a bit longer, to holding on and letting go, to settling deeply into Absorption. Like all of the other aims, mañana competence is a goal that cannot be achieved; it is a skill best cultivated over time.

12. KINAESTHETIC INTELLIGENCE

Movement intelligence supports somatic and emotional wellbeing. It is a coordinator, motivator, and regulator of daily activities, athletic performances, and everything between. Even in your sleep, it keeps you moving and adjusting to serve the needs of the body. As a twenty-four-hour coordinator, it smoothly orchestrates all of your movements, from simple to complex, unconscious to conscious, and habitual to novel. As a motivator, it prompts you to take the stairs instead of the elevator, get off the chair regularly to stretch, roll out the mat to nourish the body from within, or engage in play on a sports field. As a regulator, it takes care of the energy utilisation in your body and informs you of the extent of your healthy movement parameters. In other words, it tells you when to keep going and when to ease off.

Kinaesthetic intelligence not only makes you and keeps you functional and healthy, it also enables you to make independent body-choices. You know – from within – what is right for you at a given moment. Be encouraged! Your kinaesthetic intelligence can be either reclaimed or embraced for the first time in almost any circumstance. The more we learn about the body, the more we learn that very little is set in stone, if anything. If you feel clumsy or disconnected at the beginning, press forward. Every movement done with intention and curiosity is a step toward deepened embodiment. The movement freedom that comes with kinaesthetic intelligence is worth the energy and time investment; manifold.

IN SUM TOTAL:
6 GOOD REASONS FOR SLINGS

Envision the twelve Training Aims as a tensegrity in which one benefits and positively influences all others. The combined effects are summed up in the '6 Good Reasons for Slings', listed in the 'Slings Practice' section at the beginning of this book. Following is a more elaborate explanation of these overarching accomplishments that enhance life-quality from within.

Essence of the 12 Myofascial Training Aims

POSTURAL EASE

Adaptable postural balance and the related sense of inner equanimity facilitates a dynamically sustained physical alignment that feels restful, yet lifted and decompressive.

In the German language, posture is called "Haltung", which means attitude. Differentiated are "Geisteshaltung", the mental attitude from "Körperhaltung", the body attitude. I don't believe there is such a thing as the universal right or wrong alignment, neither for the mental nor the body attitude. Instead, it's about a personal, dynamic balance between the different aspects of our attitude(s).

In terms of posture, dynamic balance stands for adaptable postural balance. It's the kind of alignment that makes you feel at ease in the body and free to move in the ways you want or as life demands. In contrast, is a posture that is maintained by conscious muscular effort based on external ideals. Occasionally this might have its place; however, it is a short-term strategy only. The long-term aim is a postural balance dynamically sustained with the greatest degree of ease.

MOVEMENT FREEDOM

Movement poise and efficiency give the freedom to engage in desired activities and handle disfavoured or uninvited physical demands with aplomb.

Movement freedom sometimes is equated or confused with flexibility, which is commonly associated with suppleness in muscles. While muscular flexibility is important, it is only one of several elements whose dynamic interplay facilitates the poised dexterity that characterises movement freedom.

Dynamic stability: Includes balance between strength in stability-oriented, local muscles and tensile strength in associated fascia.

Global strength: Includes balanced between strength in movement-oriented, global muscles and tensile strength in associated fascia.

Limberness: Includes balance between mobility of joints, flexibility in muscles, and adaptability in fascia.

MOVEMENT LOVE

The inner motivation to be active and move with joy is health-promoting and mood-lifting, and it is the most effective medicine on the planet (with the least side effects).

We are hardwired to move because movement sustains life. Naturally, though, not everyone has the same degree of intrinsic drive or love for being active, yet every person can cultivate what they have. Like love in other forms, movement love requires attention, dedication, and the willingness to engage whole-heartedly to be sustained. This worthwhile life-long commitment supercharges your longevity with quality and makes you feel invigorated along the way.

SOMATICALLY AT HOME

The feeling of inner belonging and togetherness fosters physical, mental, and emotional equanimity.

When structurally integrated, the dynamic balance within and between the systems of the body facilitates a sense of ease that expands beyond the physical. Because the body and the brain form a unit, thought can be seen as a mental aspect of feeling and feeling as the bridge between physical sensation and emotion. In that sense, unity means every part of your physical self is mentally and emotionally integrated. Seen the other way around, you are mentally and emotionally in tune with the body in its entirety. In short, you are at home in your body.

MEANINGFUL (SELF)AWARENESS

Understanding body, mind and emotions from within facilitates integrity, authenticity and fulfilling choice-making.

Awareness is key to making satisfying and wellbeing-oriented choices. Awareness is not enough, though. One also needs to be able to draw meaning from what is perceived to understand the basis of their choices and actions.

RADIANT VITALITY

Being and feeling vibrantly alive!

It is one thing to be alive. It is another thing to feel alive, rather than merely surviving life. And it's yet another thing to be and feel vibrantly alive. Being in a dynamic state of radiant vitality or regaining it after adversity is a signifier of structural, mental, and emotional integrity.

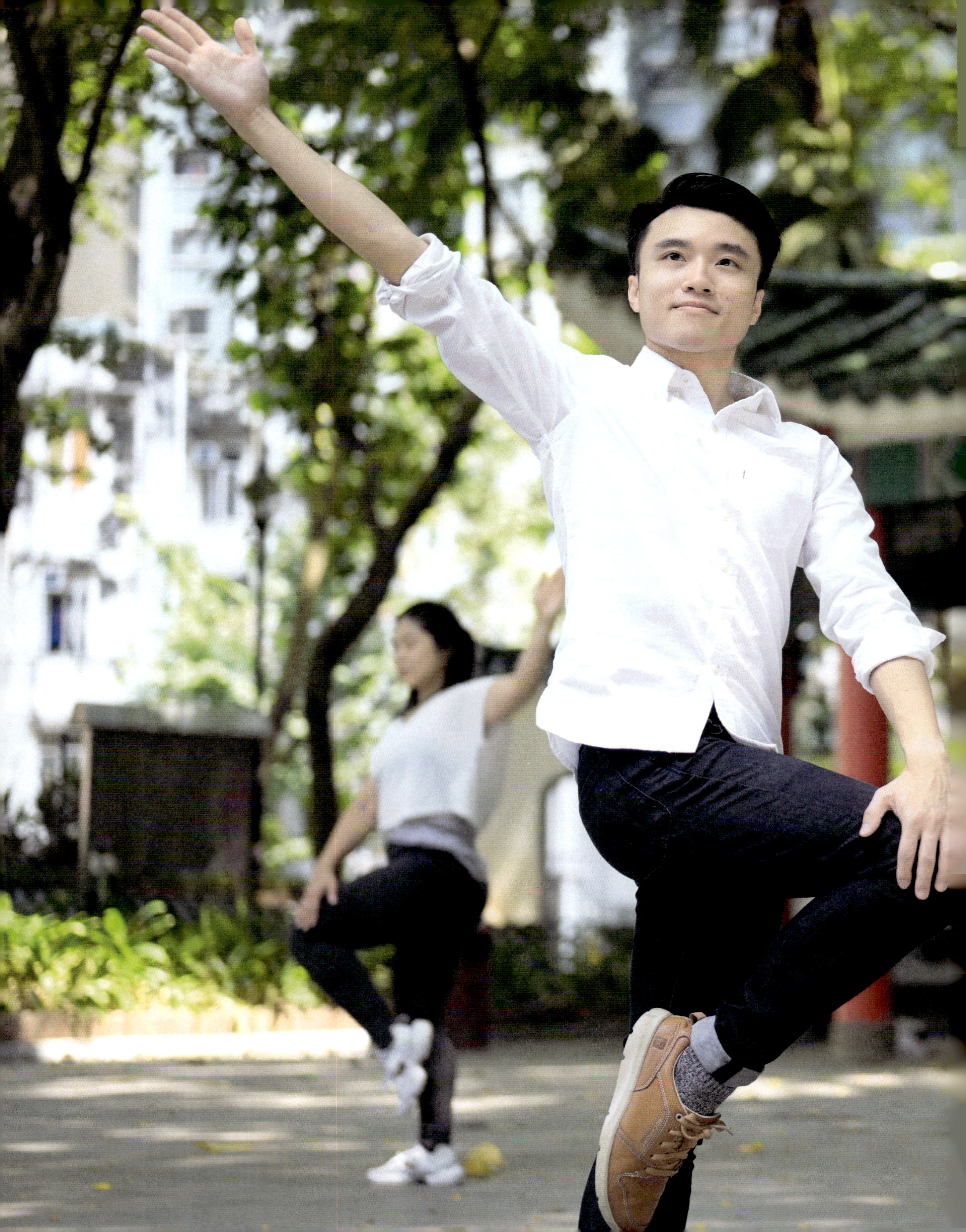

PART 3
SLINGS
MYOFASCIAL
TRAINING
APPLICATIONS

PART 3

CHAPTERS OF PART 3

8 Teaching Principles

Training Guidelines

Lesson Planning Guide

8 SLINGS IN MOTION TEACHING PRINCIPLES

Now that we have completed the exploration of the elements of the 'Slings Myofascial Training Trinity', let's discuss the Teaching Principles of a Slings lesson:

1. Fascial Movement Qualities
2. Myofascial Meridians
3. Differentiated Integration
4. Functional Choreography
5. Flow
6. (Re)Balancing on the Go
7. Resource-oriented Language
8. Authenticity

1. FASCIAL MOVEMENT QUALITIES

Careful consideration and versatile integration of the Fascial Movement Qualities

The twelve Fascial Movement Qualities described in the earlier chapters are considered in the exercise selection of a Slings lesson. To what degree individual qualities are utilised and trained is variable–deliberate ratio changes, based on the theme of a functional choreography, are encouraged.

2. MYOFASCIAL MERIDIANS

Intentional integration of all myofascial meridians

Slings lessons generally contain exercises for all of the myofascial meridians of the Anatomy Trains concept by Tom Myers. Exceptions are specialised classes that are part of a series. For example, a six-week course in which each class focusses on specific slings. Which myofascial meridians are focussed on and the specific order in which they are trained naturally varies with the theme of the functional choreography.

For detailed information about the body-minded movement aspects of Anatomy Trains, refer to the Anatomy Trains in Motion study guide and course concept.

The Lines at a Glance

1. Superficial Back Line (SBL)
2. Superficial Front Line (SFL)
3. Lateral Line (LL)
4. Spiral Line (SPL)
5. Back Functional Line (BFL)
6. Front Functional Line (FFL)
7. Ipsilateral Functional Line (IFL)
8. Superficial Front Arm Line (SFAL)
9. Deep Front Arm Line (DFAL)
10. Superficial Back Arm Line (SBAL)
11. Deep Back Arm Line (DBAL)
12. Deep Front Line (DFL)

Slings Application

Karin changing her view on integral anatomy by embodying Anatomy Trains in Motion in Berne in Switzerland

327

3. DIFFERENTIATED INTEGRATION TRAINING

One of the overarching aims of Slings Myofascial Training is improved movement freedom in everyday activities. This means being able to perform multidimensional whole-body movements with coordinated ease. Everyday functionality is essential, and something on which I believe we all can agree. Opinions may differ when it comes to 'how' to train it. Is it better to optimise complex everyday (or athletic) activities with equally complex whole-body exercises against gravity? Or is it more beneficial to consciously train individual components with differentiated movements? Before looking at the approach we take in Slings, let's have a look at what these kinds of questions indicate.

Differentiation or Integration
The questions hint that you need to decide as to which end of the training spectrum you want to position yourself and then train or teach accordingly. On one end of the spectrum is the belief that only complex multi-joint movements, preferably against gravity, lead to functional training success. At the other end of the spectrum is the idea that motility in individual joints practised in class will, without deliberate integration, morph to a coherent bodywide movement synergy outside of the classroom.

Exclusively differentiated — Either or — Exclusively integrated

Differentiated Integration

Deliberately interwoven differentiated and integrated movements

Between the two ends of the spectrum – weaving them together so to speak – lies our approach of differentiated integration.

Differentiated: Differentiated exercises deliberately train and refine individual elements of complex movements with respect to context; in a unidirectional and multidimensional manner.

Integrated: Integrated exercises progressively combine individual components to form coherent, multi-dimensional whole-body movements.

Deliberately differentiated — Both — Fully integrated

Non-Linearity

In practice, we transition between simple and complex as well as unidirectional and multidimensional movements in a non-linear fashion. Non-linearity is an important characteristic of differentiated integrated training. It means that we blend movement simplicity and complexity, unidirectional and multidimensionality motions, ease and challenge. A linear training approach is more like a one-way street in which you start with the basic exercises, progress to intermediate, and then advanced movements without going back and forth between them. In Slings, we flow with the non-linear approach, staying away from one-way streets.

Is Differentiation Really Necessary?

For some people, the question might remain as to whether movement differentiation is really necessary when everyday or athletic activities are all complex and multidimensional, even though some may seem simple or linear.

From experience, I do think differentiation is a necessary component of a practice that aims for inside-out functionality. Trying to comprehend the intricacies of multidimensional movement and train it in its entirety can be overwhelming or impossible, both for the teacher and the mover. Therefore, studying, practicing, and then gradually integrating components is not only smart but often a key to training functionally, which, by definition needs to be doable and useful.

Differentiated integration makes multidimensional movements more accessible and functional for a wider audience. It also adds quality to the practice and provides achievable stepping stones that assure an ongoing sense of progress and success.

4. FUNCTIONAL CHOREOGRAPHY

By definition, aesthetic movement relates to motion that is felt by the senses and perceived by the mind.

The term 'functional choreography' has been chosen carefully. Not because it sounds good, but because it rings true to what we do.

Functional: Something useful that serves a purpose within a specific context.

Choreography: The harmonious arrangement of moves that lead to a coherent whole.

Applying function to movement means a clearly defined task (exercise) that is part of a series of actions (seamless sequence of exercises), resulting in something of practical value (theme of the sequence). It has characteristics (exercise execution) that change in relation to context (the mover's abilities, wants, and needs).

Aesthetically expressed training for integral functionality

Functional choreography is much more than stringing together lovely exercises to create artistic movement flows. It is the skilful combination of purposeful exercises to form deliberate sequences with seamless transitions that facilitate sustained presence.

Deliberate sequence

Purposeful movements → Conscious transition → Purposeful movements → Conscious transition → Purposeful movements

performed with sustained awareness

3 Key Questions
Before putting together a functional choreography, three questions need to be asked and answered.

For whom?   Who is it meant to benefit?

For what?   What functions are meant to be improved?

How?   How are the functions improved (exercise selection, sequencing, communication, etc.)?

Functional Choreography Applied

Let me put this into a practical example for you. The chosen theme is walking because it is a function that most of us engage in daily. The functional choreography shown also adds kinetic energy to running, which increases functional value for the athletic movers in a group lesson.

Functional Choreography for Walking with Ease and Running with Spring

For whom: Movers familiar with the Slings repertoire, who are able to perform a kneeling sequence comfortably.

For what: Enhanced walking ease and running efficiency.

How: Slings exercises that focus on mobility in, glide around, and elasticity across joint movements that are essential for walking ease and running efficiency.
The exercises progress from slow to dynamic rhythms, mix linear and multidimensional movements, and dynamically change loads placed on the body. They are taught in a deliberate, yet uplifting manner.

Functional Slings Choreography

Sit Back & Curl Up

Two functional reasons:
- Segmental movement in the spine
- Alternating flexion and extension in the hip joints

Preparing for:
Dynamic Shift, 90/90 Spiralling Twist,
90/90 Dynamic Open Twist, Sit Back & Curl Up into Arch

Transition: Turning the foot out and lifting the knee into 90/90 Gate Pose

90/90 Gate Pose into 90/90 Kneeling

Two functional reasons:
- Movement coordination
- Mobility in the hip joint

Preparing for:
90/90 Dynamic Shift into 90/90 Gate Pose

Slings Application

90/90 Shift Folded

Two functional reasons:
- Mobility in the foot, ankle, and knee
- Glide across the front of the ankle

Preparing for:
Dynamic Shift, Dynamic Hip Release

Transition: Curling up into 90/90 Kneeling, followed by raising arms overhead and spiralling toward the front leg

90/90 Spiralling Twist

Two functional reasons:
- Counter-rotation of the pelvis and thoracic spine
- Elastic spiralling of the spine

Complementing:
All other exercises

90/90 Open Spiralling Twist & Arm Pendulum

Two functional reasons:
- Rotation of the pelvis over the femur
- Dynamic arm swing with spiralling of the spine

Complementing: All other exercises

90/90 Dynamic Shift

Two functional reasons:
- Elasticity in the plantar fascia, Achilles tendon, and quadriceps fascia
- Dynamic stability in the upper body during rhythmical sagittal plane movement in the lower body

Preparing for:
90/90 Dynamic Shift into 90/90 Gate Pose, Dynamic Hip Release

90/90 Dynamic Shift into 90/90 Gate Pose

Two functional reasons:
- Adaptability in the lower body
- Dynamic stability in the upper body during continuously changing alignments in the lower body

Preparing for:
Dynamic Hip Release & Side Bend

Active Hip Release

Two functional reasons:
- Active lengthening of the adductors
- Proprioceptively determining range of movement for Dynamic Hip Release

Preparing for:
Dynamic Hip Release

Dynamic Hip Release

Two functional reasons:
- Elasticity in the plantar fascia, Achilles tendon, and adductor fascia
- Dynamic stability in the upper body during multi-plane movement in the lower body

Preparing for:
Dynamic Hip Release & Side Bend

Dynamic Hip Release & Side Bend

Two functional reasons:
- Dynamically alternating adduction and abduction in the hip joint
- Elasticity across the side of the hip and upper body

Complementing: All other exercises

PART 3

Transition: Turning the externally rotated leg inward
and establishing a High Kneeling position

Sit Back & Curl Up into Arch

Two functional reasons:

- Segmental flexion and extension of the spine
- Glide in and around the quadriceps muscles, the lower back, and the abdominals

(Re)Balancing:
All unilateral exercises

Finish: Low Kneeling

The way the exercises are taught and performed deliberately changes with the skill of the movers.

5. FLOW

Everything body-mind related sounds and feels smoother with 'flow' added to it, right? With good reason, yes! 'Flow' is a delicious word, and, more than that, it holds multiple meanings for holistic movement practices. The word itself refers to abundance and proceeding from a source with smooth, continual motion. Many of us think of water streaming or fluid circulating when we envision flow. In the realms of body-minded movement, flow is also frequently associated with softness, 'flowing along', or 'going with the flow'. Although these are wonderful qualities worth cultivating, they represent only one aspect of flow. Other qualities are energy, power, and fluid resistance. This becomes evident when you think of the liquid force of ocean waves, where nature beautifully expresses its symbiosis of ease and energy.

With these diverse qualities in mind, here are four kinds of flow that are considered in Slings Myofascial Training:

1. Movement flow
2. Contained flow
3. State of flow
4. Contrasting flow

1. Movement Flow

„Kinetic melody" Oliver Sacks

Movement flow directly relates to the execution of exercises. It mainly refers to fluid clarity with which they are performed. Although all of the Fascial Movement Qualities make their contribution to movement flow, two of them are worth highlighting: glide and kinaesthetic intelligence. Fascial glide enhances movement ease and fluidity directly, at the tissue level. Gliding motions also stimulate sensory receptors. This increases proprioceptive finesse and interoceptive clarity, therefore smooth movement coordination, performed with clear intention.

Fluidly orchestrated, kinetic self-organisation

Oliver Sacks poetically coined the phrase 'kinetic melody' which we see play out beautifully in Slings as fluidly orchestrated, kinetic self-organisation.

2. Contained Flow

Thoughtfully structured movement sequences that leave creative freedom

The application of functional choreography is considered 'contained flow'. Guided movements are contrasting in their rhythm, intensity, and complexity, yet seamlessly blending and complementing each other. Transitions are as valuable as the exercises themselves, and a pause in the movement is not an interruption. Instead, it is part of the whole, like the pause in a piece of music.

Within contained flow, there is plenty of room for creativity. In fact, containment can boost creativity – both in general, as well as in lesson planning, teaching, and the practice itself.

3. State of Flow

"(Flow means) being completely involved in an activity for its own sake. The ego falls away. Time flies. Every action, movement, and thought follows inevitably from the previous one, like playing jazz. Your whole being is involved, and you're using your skills to the utmost."
Mihály Csikszentmihályi

Sustained engagement through the dynamic interplay of ease and challenge

Complete engagement or presence within an activity signifies a state of flow. When the practice engages body and mind alike, the mover is present. To stay engaged, skill and challenge need to match. If an activity is too easy, interest wanes. If it is too hard, satisfaction or motivation diminishes. The art is to find a dynamic balance between moving in and out of the 'comfort zone', 'play zone', and 'challenge zone'. It is the dynamic interplay between them that facilitates engagement and presence, and with it, the experience of a flow state.

Comfort Zone
In the comfort zone, existing skills are practised and refined. It is the feel-good realm of competency, refinement, and mastery.

Play Zone
In the play zone, existing and new movement skills are challenged in a playful manner. It is a fun realm in which physical adaptability is light-heartedly tested and boosted.

Challenge Zone
In the challenge zone, existing skills are tested, new challenges are embraced, and new skills are developed. It is the realm of humbleness, vigour, and growth, where the movement spectrum becomes broader, and the somato-psycho-emotional adaptability becomes greater.

4. Contrasting Flow

The deliberate combination of dynamic and rhythmic moves with mellow, melting motions and sustained poses assembles to the contrasting flow that is a signature quality of Slings. It can be viewed as a reflection of life's movement diversity or as a balancing complement to daily movement monotony.

Contrasting movement flow also enhances resilience, a quality that balances toughness and tenderness, both physically and emotionally.

Dynamic Movement
Periods of high intensity and contrast are incorporated to wake up and shake up physical and intellectual routines. Dynamic exercises build stamina and confidence – and if done in the right dosage, they invigorate.

Mellow Motion
Exercises that have a more steady, mellow rhythm are incorporated to promote a sense of calm within activity. Mellow motions can be a respite from the whirlwinds and stresses of daily life. A form of active meditation if you like.

Purposefully synthesised dynamic and mellow motions, melting and strong poses

6. (RE)BALANCING ON THE GO

Deliberate inclusion of (re)balancing exercises to assist change integration and a feeling of inner equilibrium

The Slings repertoire contains a wide variety of unilateral, multidimensional motions that create immediate, noticeable effects in the body. To assist the structural and sensory integration of the inner changes soon afterward, (re)balancing exercises are regularly and purposefully incorporated in the lesson.

Bilateral Movements

Flexion and extension: Bilateral flexion and extension movements are incorporated in alternation to balance and complement each other. The ratio naturally varies.

Unilateral Movements

Unilateral movements such as lateral flexion, adduction and abduction, contralateral arm and leg actions, and spiralling motions are balanced in different ways.

Contralateral motion: Unilateral movements are always executed on both sides. This is sometimes done consecutively, other times in series with other unilateral exercises.
In a one-on-one setting, the amount of repetitions on each side can vary in accordance with the rebalancing strategy outlined in a posture-based training program.

Unwinding motion: Giving the body a sense of balance before side changes can be useful in longer sequences, where a string of unilateral exercises are linked.
Also, after intense unilateral exercises, especially sustained lateral flexion or rotation of the spine, movement into the opposite direction is included immediately after. This countermovement is called 'unwinding'. A single repetition is often sufficient to facilitate a sense of ease before moving on.
The two sides of the body are still properly balanced at one point in the functional choreography.

Bilateral motion:	The two sides of unilateral exercises are usually linked and always rounded up with a bilateral movement. The transition exercise can be an insightful medium to sense somatic changes that were created by the unilateral motion. It also positively challenges our kinaesthetic intelligence to find a sense of equilibrium within an 'imbalanced state'. Concluding with a bilateral movement enhances structural balance and fosters a sense of connection between the two sides of the body.

Muscle-Focussed Attuning

Stating what is obvious by now, fascia-focussed exercises have a tremendous scope of benefits for our physical and emotional self. Many of these benefits are coupled with change, and change can be coupled with temporary uneasiness within the process. Occasionally inner life can feel chaotic, rather than dynamically balanced. To support the integration of change, stability and toning exercises have proven useful. They facilitate the dynamic stability that is structurally needed to prevent disintegration, along with a sense of inner 'togetherness'. For many of us, the latter is often just as important when going through a somatic transformation phase.

Dynamic stability:	Deep working core stabilisation exercises rebalance the temporary instability and feeling of inner chaos that can accompany fascial changes.
Toning:	The familiar activation of muscles and their robust strength can foster the reassuring feeling of inner strength and, in this way, ease the effects unsettling change processes.

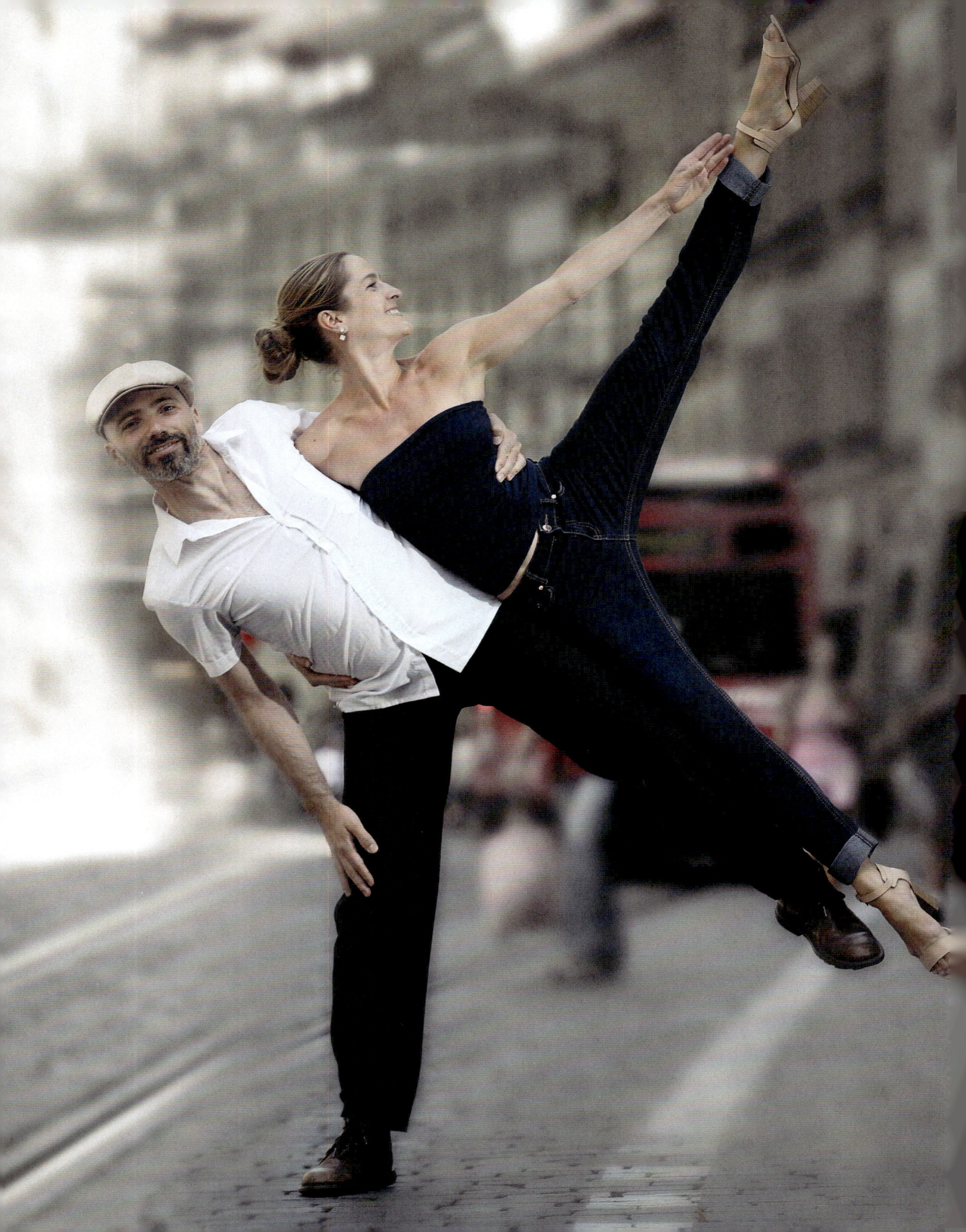

7. RESOURCE-ORIENTED, FASCIA-FOCUSSED COMMUNICATION

Language is an important component of the Slings teaching methodology. As previously discussed in the kinaesthesia section (chapter eleven of the Fascial Movement Qualities), the vocabulary we choose as teachers, influences the listener's interpretation of the characteristics of an exercise. More broadly, our voice modulation and phrasing have the power to alter meaning, while the diversity of our cueing changes how we meet the different learning styles in the group lesson.

Awareness to an exercise-coherent voice modulation, encouraging wording, fascia-focussed vocabulary, inclusive cueing

Captivating Voice Modulation

Without voice modulation, you would speak in a continuous, monotonous pitch or tone. Modulating your voice means that you adjust it to add dimension to the words. You may choose to speak louder or softer, faster or slower, leave more or less pauses between sentences, or add a specific emotional colouring to your speech. All of it contributes to the effectiveness of your communication and keeps the listener(s) engaged.

Dynamic, exercise-coherent voice modulation

Exercise-Coherent Dynamic Balance

Sometimes in the realms of body-minded movement, the voice of a teacher can become somewhat flavourless, when it is meant to be soothing. While this is well-intended, too much serenity in the voice intonation can be soporific. On the other side of the spectrum is the animator, whose voice is always peppy and a tad too loud in volume and vocabulary. Ideally, the diversity of the Slings repertoire is reflected in the dynamically balanced voice modulation, the wording, and body language of a teacher.

Resource-Oriented Communication

The language we use in class – and ideally for ourselves as well – is resource-oriented. To have a clearer image of what this means, it is useful to compare a resource-oriented form of communication with other forms, including one example for each.

Sincerely encouraging wording

343

Other Forms

Negatively phrased: Don't move your pelvis when peeling the foot off the floor.

Dysfunction-focussed: Keep your abdominals firmly engaged, because arching your back off the floor creates unhealthy compression, which can hurt your back. If you feel pain, place both feet on the floor to reestablish a centred alignment.

Risk-focussed: If you have a history of knee pain, be careful with this movement as it might trigger discomfort.

Presumptuous: When sitting in a centred alignment with the legs extended is challenging, your hamstrings are tight.

Issue-focussed: Sitting with your pelvis tilted back makes you slouch in your lower back, while the aim is to lengthen the spine.

Judgmental: The aim of this exercise is to lengthen the spine during the side bend. Collapsing is a sign of poor strength or movement coordination.

Competitive: Some of you are executing this exercise very well, while others seem to be stagnating in their progress.

Habitual: Well done, everyone, as always.

Derogative: In contrast to Yoga/Pilates/fitness/other modalities, we execute spinal rotation in a more functional/healthier/sensible way.

Resource-Oriented

Neutrally phrased:	Keep the pelvis centred while peeling the foot off the floor.
Function-focussed:	Keep your abdominals firmly engaged and pay special attention to the length in your lower back. Feeling your back muscles work is natural and healthy. When you notice your back arching off the floor, place both feet on the mat to reestablish a centred alignment.
Wellbeing-focussed:	If you have a history of knee pain, tune in and listen to your body. Let your knees tell you what movement dosage is right for them today.
Informative:	When sitting in a centred alignment with the legs extended is challenging, it can have a variety of reasons, which we will explore later in class.
Solution-focussed:	For now, to achieve the desired length in your spine, feel free to bend the knees or sit on a cushion.
Non-judgmental:	The aim of this exercise is to lengthen the spine during the side bend. Moving to the end range is halfway too far; lift the bottom ribcage to regain spaciousness in the waistline.
Progressive:	How do you feel with your own progress in this exercise? If you feel satisfied that is great. If you feel 'held back' today, notice where and how, so you can create gradual change as you move forward.
Current and honest:	The movement control was excellent today, well done. Regarding rhythmicality, there is potential for more buoyancy. Focus on letting the arm swing freely, and the body follow effortlessly for the next few repetitions.
Respectful:	The way we spiral the spine in Slings might be different than what you have experienced before. Enjoy the difference!

Fascia-Focussed Communication

To facilitate a fascia-focussed and, therefore, a different way of embodying known and novel exercises requires a new kind of vocabulary, I believe. Experience has shown that teaching new things in old ways can slow down the embodiment process.

Ideally, the new vocabulary we adopt to describe Slings exercises represents the Fascial Movement Qualities they utilise. The modulation of the voice adds an affective, motivational dimension to the information. At their best, the words and voice intonation echo characteristics of Fascial Movement Qualities, applied Training Techniques, and aspired Training Aims.

When they do, the clear description of an exercise focussing on tensile strength makes us feel fascially strong, yet light and expansive. A hearty muscle-focussed cue fosters a sense of solid inner strength, while springy bounces expressed in an upbeat voice facilitate the buoyancy that comes with fascial elasticity. This is contrasted by the soothing quality of a low, slow voice that can soften inner resistances, so we can yield into a melting pose.

Diverse Cueing

Different people have different learning styles, which go hand in hand with the way information is taken in and processed. To communicate effectively and reach, therefore, benefit as many people as possible, we blend a broad variety of instructions in Slings lessons. The four styles of cueing listed below can convey the same kind of information in very distinct ways. It enables us to meet individuals in their preferred communication style some of the time, while at other times, they expand their scope of learning and embodiment by translating unfamiliar or non-preferred instructions into motion.

Training oneself in diverse cueing is just as beneficial for us teachers because we too have our preferences. Deliberately describing exercises in a less familiar form extends our own panorama of communication skills; in front of a class and in everyday life.

4 Distinct Cueing Styles

Technical instructions:	Describe the movement execution from gross to detail. *Proprioceptively-focussed*
Kinaesthetic instructions:	Describe feelings, encourage to notice feelings, and draw meaning from what is felt. *Interoceptively-focussed*
Imagery:	Uses images and encourages visualisation to feel the body and movement differently. *Interoceptively-focussed*
Tactile instructions:	Utilise touch to cue body positions, guide movement, and convey specific feelings. *Proprioceptively and interoceptively-focussed*

The ratio between the different cueing styles varies depending on the participants, the degree of embodiment of the exercises, the theme of the lesson, and the feel of the day.

8. AUTHENTICITY

The eighth Teaching Principle sums up the heart of the matter: be yourself.

Remember the fifth Guiding Principle of Slings: 'Diversity-Embracing'. The Slings concept is intended to bring out the best in you personally and let your authentic self shine; in front of a class, if you are a teacher, on the mat as a mover, and in everyday life as a unique and good-as-you-are human being.

Be your most resourceful, authentic self and let the world mirror your attitude.

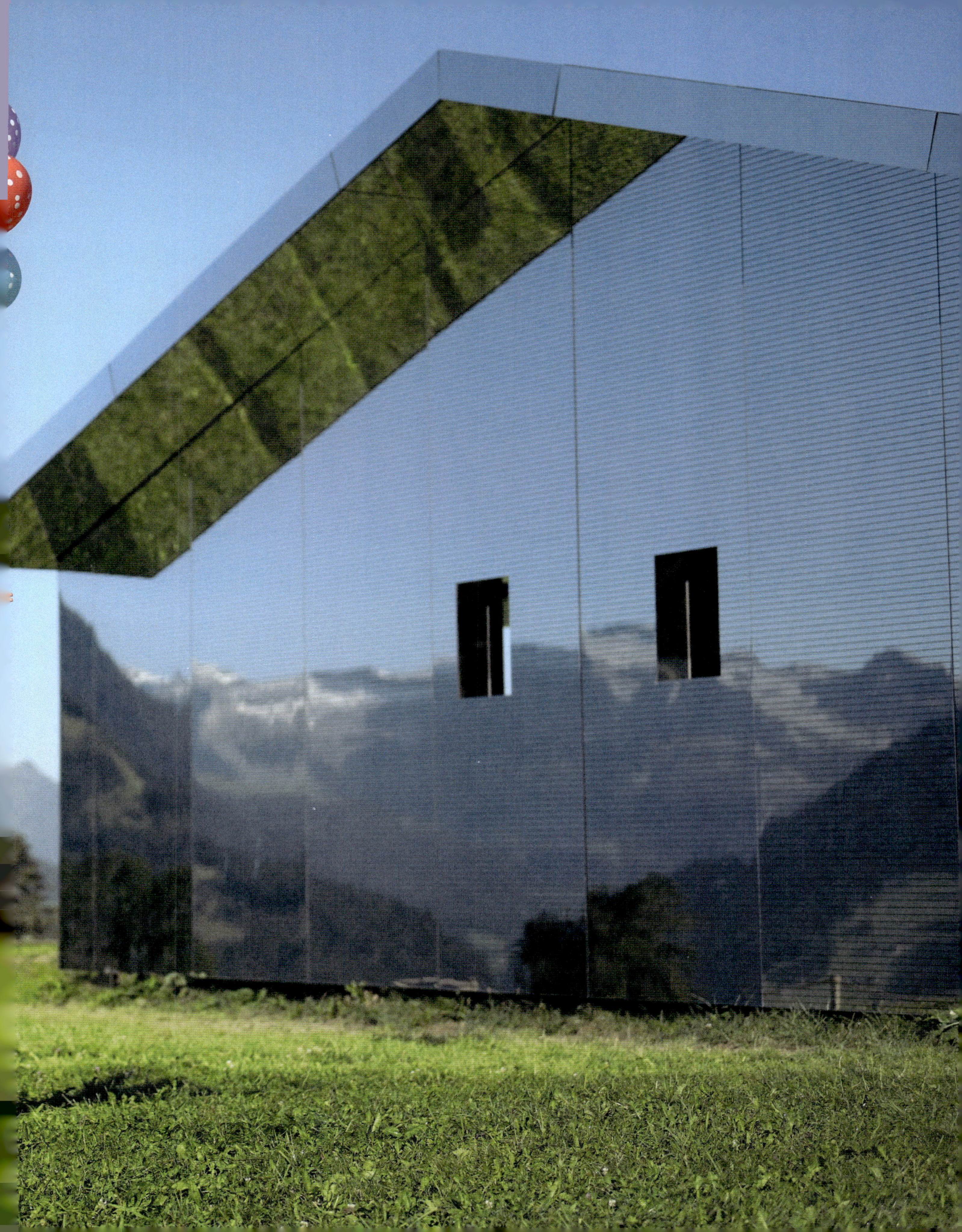

PART 3

SLINGS LESSON PLANNING GUIDE

"Skilful movement is a reminder of the joy of exercise, but also of the discipline and effort required to do it well."
— Damon Young

In terms of lesson planning, there is no such thing as a 'one size fits all' sequencing of exercises. There is, however, a lesson planning format, which has proven successful in practice. It considers the layering of the myofascial meridians and their functional aspects to facilitate the twelve Slings Training Aims.

The following guide is for planning a 60-minute lesson. It can be easily extended to 75 or 90 minutes.

Slings Application

Theme	Timing	Direction	Body Focus	MM(s): Primary	MM(s): Assisting	Energetic Quality	Primary Movement Dimension(s)	Primary Exercise Type	Included Primary Movements
1. Arriving: Getting moving	15-20	Outside in	Feeling movement	Various	Various	Awakening	Sagittal plane body motion, multidimensional arm movements	Mobilising	Centered body alignment and/or trunk and leg flexion Multidimensional arm and shoulder movements
2. Centering: Sense of balance		Inside out	Core awareness	DFL	SBL	Grounding	Stillness	Core activation	Centered body alignment Expansive breath
3. Easy flow: Warming up		Outside in	Global pliability	SBL	DFL, SFL	Gently energising	Sagittal plane slow motion	Mobilising, active lengthening, strengthening	Segmental spine flexion Hip flexion Dorsiflexion and plantar flexion
				LL, ALs	DFL		Frontal plane slow motion		Lateral spine flexion Generous arm and shoulder movement
				SFL	DFL, SBL		Sagittal plane slow motion		Segmental spine extension Hip extension
4. Contrasting flow: Body of lesson	25-30	Inside out Outside in	Expanding movement scope and diversity	SPL DFL, LL FLs SBL, SFL	DFL, ALs ALs DFL, ALs DFL	Energising, challenging, satisfying, grounding	Multidimensional motion in all planes, at different rhythms, and with various intensities	Diverse blend of rhythmical dynamic exercises, grounding poses, and mellow motions	Whole-body movement including three-dimensional motion of the: • Spine, including spiralling • Hip joints, including rotation of the pelvis over the femur • Shoulders, including movement of stabilised scapula over the humerus Dynamic stabilisation of the: • Spine • Pelvis • Shoulder girdle and shoulder joint
5. Gentle flow: Winding down	5-10	Inside out	Rebalancing	SBL, SFL DFL	ALs	Calming	Sagittal plane slow motion	Gentle mobilising, balanced lengthening and toning, core stabilisation	Flexion and extension of the: • Spine • Hip joints • Knees • Feet Dynamic stabilisation of the: • Spine • Pelvis
6. Absorption & Invigoration: Feeling relaxed and awake	5	Inside	Stillness	DFL		Rejuvenating, inward focus	Stillness		Most comfortable supine position
		Inside out	Awakening	SBL, SFL SPL SFL, SBL	DFL DFL, ALs DFL, ALs	Gently vitalising, opening outward	Sagittal plane and transverse plane motion	Back massage followed by gentle spinal spiralling followed by upper body flexion, extension, centering	Spine flexion Gentle, multidimensional spinal movement Spine flexion, extension, and centering

SLINGS TRAINING GUIDELINES

As a general rule, regular, short practice sessions that utilise various fascial movement qualities are preferred to irregular, long practice sessions that focus on only one or two fascial movement qualities.

Daily Training Frequency

Because Slings Myofascial Training utilises a broad spectrum of Fascial Movement Qualities and consists of a diverse blend of repertoire, there is no system overload. It can be practised every day. Realistically though, very few people prioritise their movement practice in a way that makes it a daily routine. The good news is that less is just as valuable and brings an abundance of short and long-term benefits. It really is up to the individual to create a sustainable training schedule that is coherent with their everyday commitments.

Training Duration and Weekly Frequency

The duration of one Slings session can range from ten to ninety minutes. That means, either short intervals several times a week or longer lessons at least once or twice a week.

Training Span

Although there are notable effects within a few weeks of training, remember that fascia is a slow renewal system, and rebuilding fascial architecture takes time. It means that fascia-focussed training is a longstanding commitment with prompt and long-term benefits. Create a sustainable schedule that matches the rest of your life, say 'yes' from within, and stay with it; it is worth it!

Slings Application

Training Schedule

An hourly Slings lesson once a week is 100% better than none. Just keep in mind that the training frequency is a choice in which the positive outcomes match the investment.

	Duration	Frequency
Slings Light	10-15 minutes	4 to 5 times a week
Slings Regular	30 minutes	3 to 4 times a week
Slings Regular	60 minutes	2 to 3 times a week
Slings Pro	30 to 60 minutes	4 to 7 times a week

Mix and Match

Feel free to mix and match differently timed sessions as you like. The following are some examples of personalised training schedules.

	Mo	TU	WE	TH	FR	SA	SU
Slings Light	15 min.		15 min.		15 min.		30 min.
Slings Light	15 min.	30 min.		15 min.		30 min.	
Slings Regular	30 min.	15 min.		60 min.	15 min.	30 min.	
Slings Regular	60 min.		30 min.		30 min.		60 min.
Slings Pro	60 min.	15 min.	30 min		30 min.	60 min.	

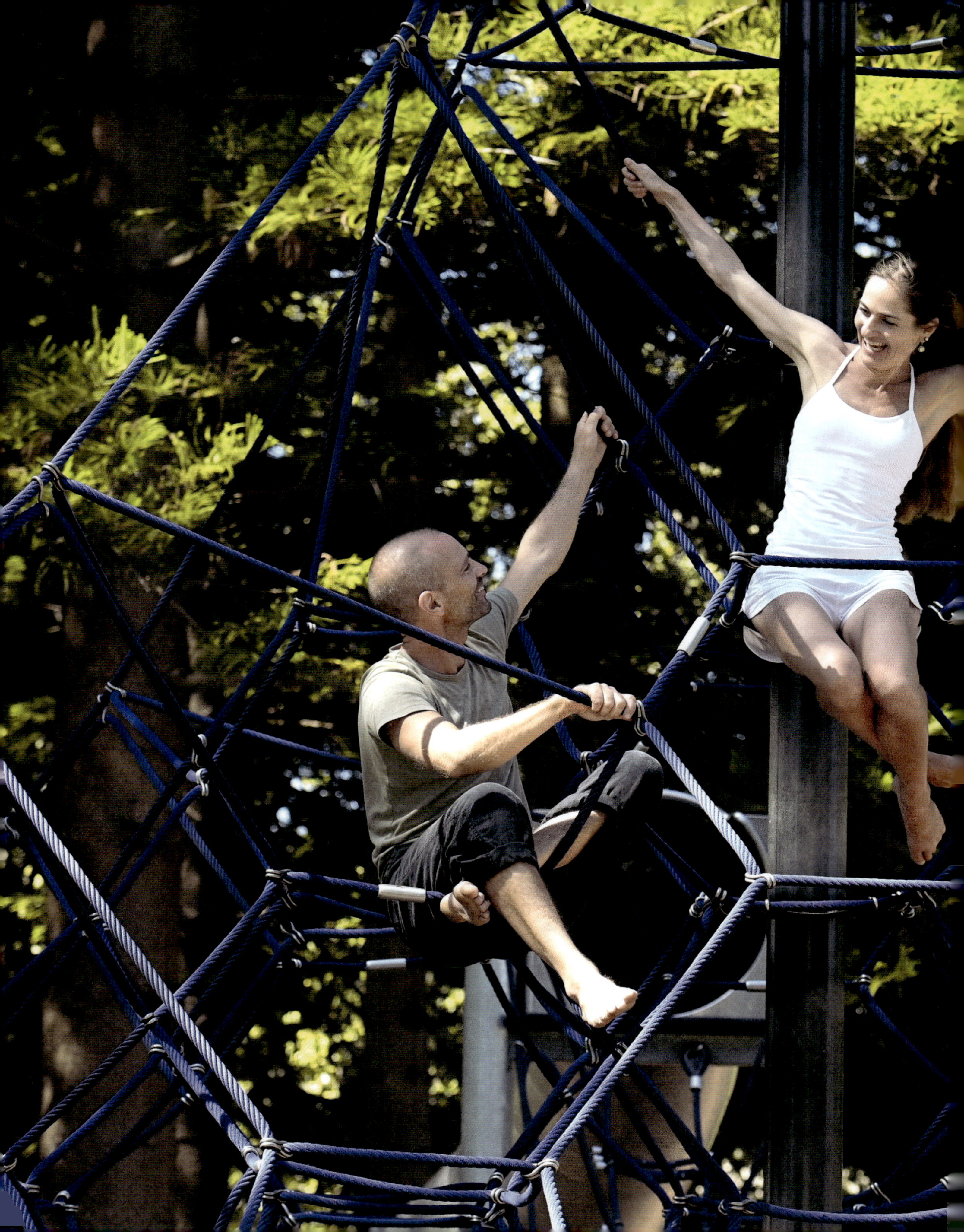

PART 4
PARTS OF THE WHOLE AND THE WHOLE OF THE PARTS

PART 4

CHAPTERS OF PART 4

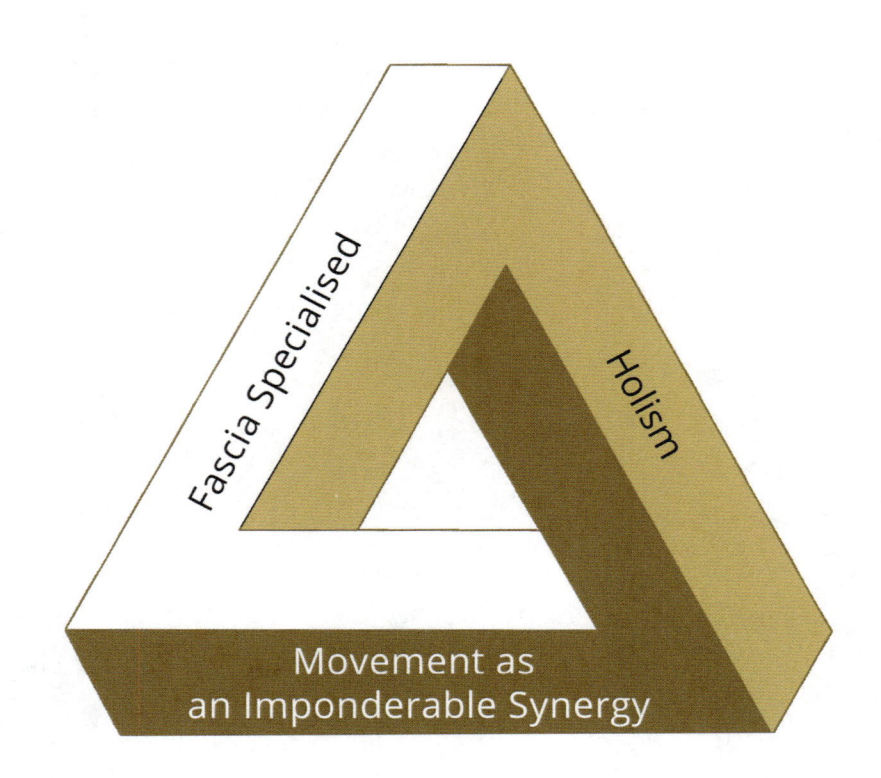

Fascia Specialised

Holism

Movement as
an Imponderable Synergy

Parts of the Whole

4 FASCIAL TYPES

SUPERFICIAL FASCIA

The fascia Gil Hedley once called the 'wedding suit'.

The sensitive outer shaper

Carmen and Marc in figure-hugging and figure-flattering attire on their special day in a castle in Switzerland

The superficial fascia is between the skin and the first layer of deep fascia; or more accurately, beneath the two layers of the skin, the epidermis and dermis, and above the aponeurotic fascia.

Functionally speaking and when referring to superficial fascia, the dense fascial membranes and adipose tissue beneath the skin are implied. Like a warming neoprene wetsuit, the highly adaptable superficial fascia hugs and shapes our body. Despite its pliability it has noteworthy force transmitting tensile strength. Naturally, it is more present in some areas and fuller in some people. Wherever and in whatever form it presents itself, superficial fascia should be appreciated, moved, and treated with care.

Architectural and Functional Characteristics

- The superficial fascia contributes to the uniqueness of your body's shape and, therefore, your appearance (in a special way, like a wedding gown on that special day).
- Together with the skin, it forms a protective and insulating bodysuit.
- The superficial fascia can be more loosely or firmly connected to the skin and deep fascia.
- Together with the skin, the superficial fascia slides over the underlying deep fascia.
- The fibre orientation is multidirectional.
- Superficial fascia has tensile strength, yet is highly adaptable and moveable in all directions.
- Its loose texture accommodates the tortuosity of blood vessels and nerves. In turn, the sinuosity of blood vessels and nerves within the superficial fascia permits the tissue's great adaptability.
- It features channels for lymphatic flow.
- There superficial fascia is highly innervated, therefore kinaesthetic.

SUPERFICIAL FASCIA IN MOTION

"Superficial fascia takes a whisper to change."   Gil Hedley

The superficial fascia deserves as much attention in body-minded, fascia-focussed movement training as the acclaimed deep fascia and the loose fascia that has gained stardom more recently. Through self-massage exercises and generous movements, we aim to enliven the superficial fascia, maintain its pliability, and support fluid flow within.

As a reference, the Fascial Movement Qualities and Training Techniques in practical focus are listed below. Gold font shows what is in the centre of attention, black font implies relevance without strong focus, and grey font indicates no immediate focus in the practical applications. As always, the selections are interpretive to a degree, therefore intended to be a guide, not 'the law'.

12 Fascial Movement Qualities
Utilised:

1. Tensile Strength
2. Muscle Collaboration
3. Force Transmission
4. **Adaptability**
5. **Multidimensionality**
6. Fluidity
7. Glide
8. Elasticity
9. Plasticity
10. Tone Regulation
11. **Kinaesthesia**
12. Imponderability

12 Training Aims
Applied:

1. Stabilising
2. Toning
3. Pushing, Pulling, Counter-traction
4. **Active Lengthening, Expansion**
5. **Spiralling**
6. **Hydrating**
7. **Gliding**
8. **Domino Motion**
9. Bouncing and Swinging
10. **Massaging**
11. Active Ease
12. Melting and Invigorating

Mermaid with Double Spiralling in Z-Sit on Massage Ball

PART 4

LOOSE FASCIA
The ugly duckling that turned into a luminous swan.

The integral glider

Image by Pezibear from Pixabay.com

The loose fascia has been the 'ugly duckling' of anatomy, overlooked, disregarded, and scraped away. Applause for the anatomical visionaries whose 'jamais vu' capacity (the ability to see known things in new ways) changed our view of this sensory-rich, fluid tissue.

Unrecognised as it was, loose fascia didn't even get a name when anatomical nomenclature was defined. Because of its anonymity, it adopted several different names in more recent times. The terms range from endearingly descriptive to anatomically oriented. In his memorable 'Fuzz Speech' Gil Hedley called it just that, 'the fuzz'. Now he identifies it as 'perifascia'. Jean-Claude Guimberteau uses the term 'sliding system', others have opted for 'intermuscular fascia'. The most recent and probably most celebrated name is the 'interstitium'. We chose the descriptive and hopefully memorable term 'loose fascia'.

Architectural and Functional Characteristics

Simply put, it can be described as the multidirectional, watery tissue that surrounds, links, and enables stabilised glide within muscles, between fascial structures (including neighbouring epimysia, therefore, between muscles), and other organs of the body.

- Loose fascia is found throughout the body.
- It is watery and highly adaptable.
- The fibre orientation is multidirectional.
- Loose fascia connects neighbouring structures in a flexible manner.
- It forms a sliding surface between myofascial structures.
- Loose fascia is highly innervated, therefore kinaesthetic.
- The gliding movements stimulate sensory receptors.

LOOSE FASCIA IN MOTION

In Slings, we aim to stimulate, nourish, and generously move loose fascia to maintain the tissue's natural fluidity and multidimensional sliding ability. Unrestricted glide is promoted wherever it is or feels functionally needed.

12 Fascial Movement Qualities
Utilised:

1. Tensile Strength
2. Muscle Collaboration
3. Force Transmission
4. **Adaptability**
5. **Multidimensionality**
6. **Fluidity**
7. **Glide**
8. Elasticity
9. Plasticity
10. Tone Regulation
11. **Kinaesthesia**
12. Imponderability

12 Training Aims
Applied:

1. Stabilising
2. Toning
3. Pushing, Pulling, Counter-traction
4. **Active Lengthening, Expansion**
5. **Spiralling**
6. **Hydrating**
7. **Gliding**
8. **Domino Motion**
9. Bouncing and Swinging
10. Massaging
11. Active Ease
12. **Melting and Invigorating**

90/90 Side Stretch with Spiralling

Parts of the Whole

DEEP FASCIA
The distinguished superstar of fascia.

The acclaimed multitasker

Jenny starring on stage in Berne in Switzerland

Deep fascia is often referred to as 'fascia profunda'. Profunda means deep in Latin. This much-celebrated part of the fascial system is highly collagenous and well-organised. It permeates, encases, and connects muscles and bones. Included in the term deep fascia are:

- Dense, planar tissue sheaths like septa (partitions), aponeuroses (tensile sheets), retinacula (stabilising, proprioceptive retainers), and joint capsules (sleeves around joints).
- Local densifications like tendons or ligaments.

The muscle fascia, namely the epimysium, perimysium, and endomysium, is also under the banner of deep fascia. However, for functional reasons and training specificity, it is given its own category in Slings.

Architectural and Functional Characteristics

- Deep fascia contains a high ratio of collagen fibres, which gives the tissue stability and tensile strength.
- Commonly, the collagen architecture is clearly structured.
- Because of the versatile functions and structural relationships of the deep fascia, the architectural patterns of the collagen fibres vary greatly. The fibre orientation can be primarily unidirectional, like in tendons and ligaments, or multidirectional, as seen in joint capsules.
- Density and elasticity are also highly variable. The degree of elasticity depends on the tissue composition (ratio of collagen and elastin fibres) and collagen architecture.
- The innervation of deep fascial structures varies depending on function.

90/90 Lunge & Side Bend

90/90 Shift Folded

Inverted V & Dynamic Double Knee Bend with Jump

Parts of the Whole

DEEP FASCIA IN MOTION

Deep fascia is intimately linked with muscles and bones, both fascially and functionally. Because of its inherently diverse characteristics, in Slings, we intentionally engage the deep fascia in a versatile manner by utilising a broad range of exercises. The repertoire includes slow and steady, rhythmical and dynamic, linear and multidimensional movements, and everything between.

12 Fascial Movement Qualities
Utilised:

1. **Tensile Strength**
2. **Muscle Collaboration**
3. **Force Transmission**
4. **Adaptability**
5. **Multidimensionality**
6. Fluidity
7. Glide
8. **Elasticity**
9. Plasticity
10. **Tone Regulation**
11. **Kinaesthesia**
12. Imponderability

12 Training Aims
Applied:

1. **Stabilising**
2. **Toning**
3. **Pushing, Pulling, Counter-traction**
4. **Active Lengthening, Expansion**
5. **Spiralling**
6. **Hydrating**
7. Gliding
8. **Domino Motion**
9. **Bouncing and Swinging**
10. **Massaging**
11. Active Ease
12. Melting and Invigorating

Roll Down

MUSCLE FASCIA

The fascia in which muscles are interwoven.

Muscle fascia sometimes spelled muscle fasciae (plural), is also known as intramuscular fascia and myofascia. It is commonly viewed as deep fascia in which muscle fibres are woven in. Because the fibrous architecture and relationship with muscles differ from those of the previously named deep fascial structures, the muscle fascia is given its own category in Slings.

Muscle-focussed thinking is deeply imprinted in the minds of those of us trained in standard anatomy. Remind yourself regularly that a muscle is not only made up of muscle tissue but the fibrous connective tissue that gives it integrity and supports smooth functioning.

Parts of the Whole

The collaborative inner sculptor

Karin sculpting her muscles from within on Grimsel Glacier in Switzerland

Architectural and Functional Characteristics

- The epimysium, perimysium, and endomysium make up the muscle fascia.
- The fibrous architecture is multidirectional, forming stabilising yet adaptable double lattice patterns.
- Muscle fascia is both force transmitting and gliding.
- Embedded in the muscle fascia are the muscle fibres.
- Muscle fascia and muscle fibres form what we call a muscle.

CHANGE YOUR VIEW ABOUT FASCIAL ARCHITECTURE

Muscle fascia is an illustrative example of how viewing anatomy from a fascial perspective differs from the traditional outlook.

Here is a standard description of muscle fascia:

- Endomysium is the delicate connective tissue that surrounds individual muscular fibres.
- Perimysium is the connective tissue sheath that surrounds bundles of muscle fibres.
- Epimysium is fibrous connective tissue that surrounds each muscle.

Viewing myofascial relationships in this way, the muscle fascia is seen as individual layers that surround muscle fibres, muscle fibre bundles, and the muscle itself, until they merge to form a tendon.

The fascial perspective is very different. Envision a piece of honeycomb and within each compartment a smaller honeycomb structure, and within each compartment again, another honeycomb. And now translate this visualisation to muscle fascia:

- Epimysium is part of an interconnected honeycomb-like fascial web in which muscles are embedded.
- Perimysium is a honeycomb-like network that encases muscle fibre bundles.
- Endomysium forms another honeycomb-like structure. Inserted into its compartments are the muscle fibres.

When looking at the relationship between muscles and muscle fascia from this perspective, fascia is recognised as a continuous web in which muscle tissue is woven in, not a series of discontinuous layers that serve as wrapping.

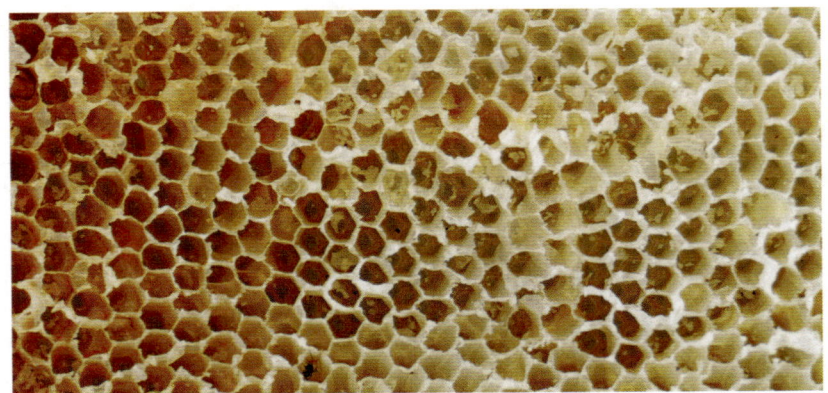

Image by Miss Suki from Pixabay.com

MUSCLE FASCIA IN MOTION

Because of the intimate relationship between muscle fibres and muscle fascia, they affect one another directly and immediately. A muscle contraction, conscious or not, creates a tensional response in the surrounding fascial network. At the same time, tensional changes and sliding motions in the muscle fascia stimulate the muscle tissue within.

When considering the coexistence of muscle contraction and fascial tension, muscle fascia can be viewed as a 'strength amplifier'. Simply put this means, that when a muscle contracts, the muscle fascia is tensioned. When the tensional end range is reached, the firmness in the fascial encasement consolidates the strength generated by the muscle within. The force is then transferred to the tendons, ligaments, periostea, and the bones themselves, resulting in efficient joint stability or movement. Force transmission is especially prominent in the epimysium and endomysium, while the perimysium effectively functions as a gliding structure.

By varying the direction, intensity, and rhythm of the exercises, we aim to strike the balance between multidimensional muscle-focussed and fascia-focussed motions to achieve the main training aim: dynamic balance and a functional interplay between muscles and muscle fascia throughout the body.

12 Fascial Movement Qualities
Utilised:

1. **Tensile Strength**
2. **Muscle Collaboration**
3. **Force Transmission**
4. **Adaptability**
5. **Multidimensionality**
6. **Fluidity**
7. **Glide**
8. **Elasticity**
9. **Plasticity**
10. Tone Regulation
11. **Kinaesthesia**
12. Imponderability

12 Training Aims
Applied:

1. **Stabilising**
2. **Toning**
3. **Pushing, Pulling, Counter-traction**
4. **Active Lengthening, Expansion**
5. **Spiralling**
6. **Hydrating**
7. **Gliding**
8. Domino Motion
9. Bouncing and Swinging
10. **Massaging**
11. **Active Ease**
12. **Melting and Invigorating**

PART 4

Twisted Pelvic Curl

Twisted Shoulder Bridge & Leg Lift   Transition

Transition   Easy Twist

Transition

Pelvic Curl

Diamond Legs

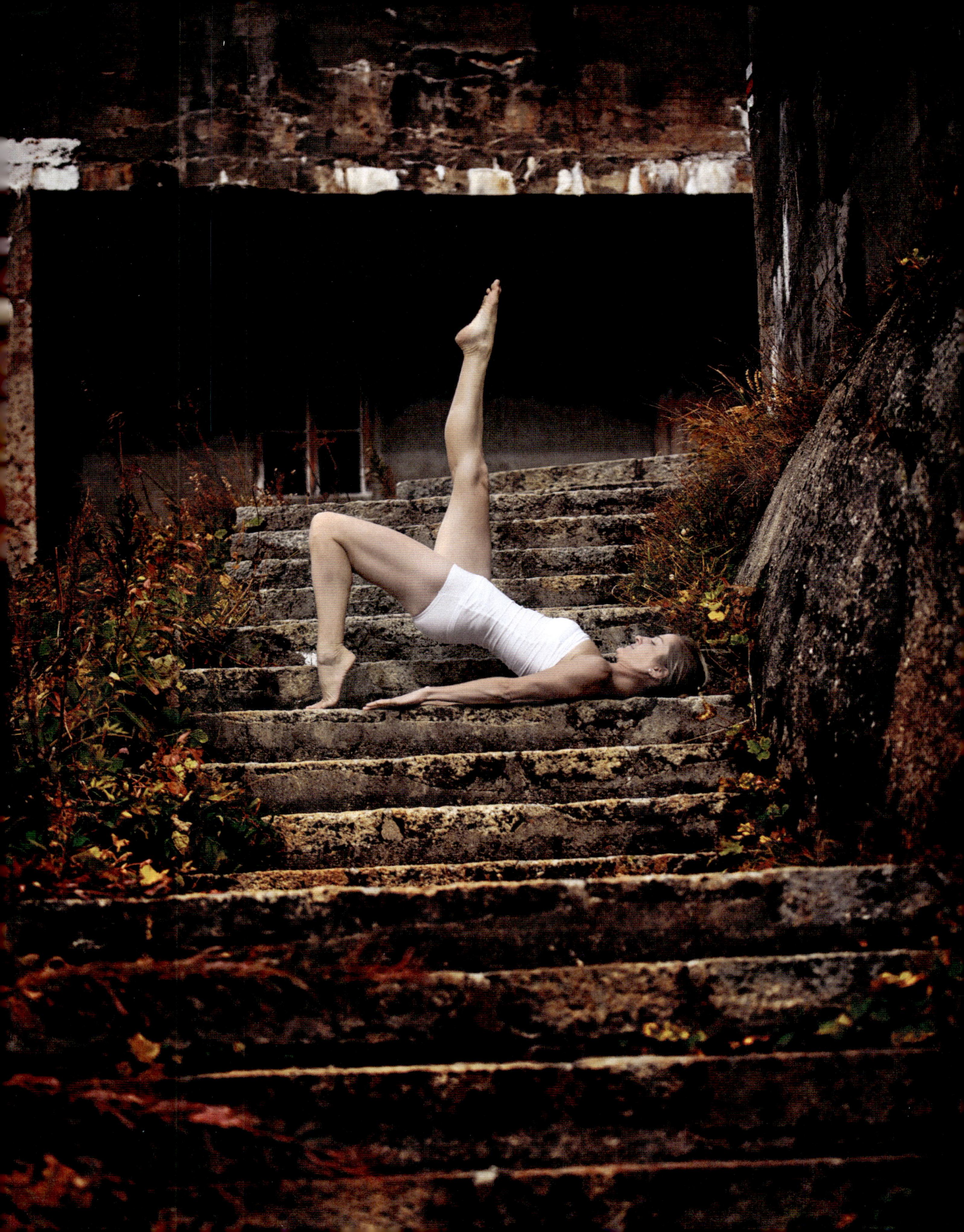

MOVEMENT AS A SYNERGY

"The heart has its reasons of which reason knows nothing". Blaise Pascal

What Pascal, a philosopher of the seventeenth century, calls the heart, is the faculty that makes us know things by sentiment, instinct, and intuition. With his quote, he says that you need the heart to fathom the imponderable. Reason or our intellect is oftentimes unable to grasp what the body intrinsically understands. The following definition of what movement is aims to marry somatic wisdom and cognitive understanding.

First and foremost, it is important to recognise that movement is not a top-down event in which the nervous system coordinates the myofascial system, which moves bones accordingly. Instead, movement is a synergy with tensegral qualities in which the physical interacts with mental and affective processes. By taking the liberty to change Pascal's quote we could say "movement has its reasons of which the analytical part of the brain knows nothing".

Movement is a neuro-myo-fascial-skeletal-psycho-emotional-perceptual synergy that is linguistically and socially influenced and in its wholeness imponderable

My current definition of movement describes it as neuro-myo-fascial-skeletal-psycho-emotional-perceptual synergy that is linguistically and socially influenced and in its wholeness imponderable. Rolls off the tongue smoothly, doesn't it? Let me explain what I mean in simple terms.

A Word About Imponderability

By definition, the word 'ponderable' means "that something can be considered carefully or deeply" and that it "can be weighted". Interestingly, the word 'imponderable' is commonly defined as "an entity or force that cannot be precisely determined, measured, or evaluated". The interesting part is that the imponderable is still considered ponderable, just not exactly. For me, 'imponderable' stands for what is so vast and complex that cognitive comprehension of all its aspects and what they mean in relation to the whole is beyond my current intellectual capacity.

Neuro: The nervous system proprioceptively coordinates and interoceptively motivates movement.
In turn, the way we move is registered and responded to by the nervous system, unconsciously to keep us healthy, consciously to function successfully in a purpose-oriented manner.

Myo: Muscles are the reflexive and deliberate movement executors.
In addition to their great physical effects, the state of muscles noticeably influences the way we think, feel, and perceive ourselves and those around us, as seen in gyms and on covers of health magazines around the world.

Fascial: With all its movement-related qualities, fascia facilitates postural ease, movement freedom, and the finest personal performance sustainably.
Structurally and functionally, it is inseparably linked with the nervous, muscular, and skeletal systems, all of which feed body-related information back to the fascial system.

Skeletal: As levers, bones add range to movement.
They also act as structural supporters and function as spacers in the body. By 'pushing out' from within, they aid healthy tension in the fascial system, which in turn provides a spacious suspension system for bones and muscles.

Psycho: Thoughts change movement intention, habitually or spontaneously.
Sometimes it is long-held beliefs, other times current ideas or a random thought that shape the way we stand and move, from Tadasana to 'holding the head up high', to offering an inviting gesture or questioning eyebrow raise.

Emotional: Feelings modulate movement in a regulatory and expressive manner.
The way we consciously or unconsciously experience affective states changes the way we move. Affective states include the whole kaleidoscope of human emotions, ranging from fear, grief, anger, disappointment, and disgust to happiness, contentment, trust, optimism, wonder, and gratitude. Based on the way we feel, the proprioceptive sense orchestrates our body language to mirror our emotional states. At the same time, the interoceptive sense motivates movement adaptations geared toward maintaining a feel-good state or regaining equilibrium. This kinaesthetic reorganisation successively alters the way we feel.

Perceptible: The meaning we draw from thoughts and feelings reshapes movement in the most personal way.
In turn, the way we stand and act changes our self-perception and how we interpret our experience of the outside world.

Linguistical: The language we use and hear (internal or spoken), and the intention with which the words are expressed, alters the message and how it is interpreted. Words have the power to transform our postural and movement patterns right now and long into the future. Vice versa, our body language impacts our speech, from voice modulation to the choice of our vocabulary.

Social: Our immediate social and extended living environment also affects the way we habitually or temporarily move.
Postural and movement patterns differ in amicable and hostile environments; social norms and values take their effect, so does our desire to move 'with or against the stream'.
In turn, our body language contributes to the dynamics within our immediate and even extended social surroundings.

Imponderable: Considering all of the above and the elements left unconsidered, movement in its wholeness is imponderable.
Acknowledging that wholeness cannot be understood by the sum of its parts fosters a healthy sense of humbleness and hopefully, curiosity. It also serves as a reminder that "the heart has its reasons of which reason knows nothing".

Correlations

In holistic systems, all elements correlate. Meaning, all elements influence each other, and as one element changes its behaviour, it creates change in the whole system.

Neuro: The nervous system coordinates movement.
Movement feeds information to the nervous system.

Myo: Muscles modulate movement.
Movement develops and sustains the health of muscles.

Fascial: Fascia facilitates ease and efficiency of movement.
Movement sustains the health and functionality of fascia.

Skeletal: Bones enable the outer expression of movement.
Movement strengthens bones.

Psycho: Thoughts influence the execution of movement.
Movement influences the processing of information.

Emotional: Feelings shape the quality and expression of movement.
Movement alters the experience of sensations and emotions.

Perceptible: Perception changes the way we think and feel about movement.
Movement influences the meaning we draw from thoughts and feelings.

Linguistical: Language alters movement expression.
Movement changes with the way we express ourselves.

Social: Immediate and extended environments impact on movement behaviour.
Body language influences social interactions.

Body-minded, fascia-focused movement makes the imponderable somatically tangible.

PART 4

THE PROBLEM WITH HOLISM

"As we acquire more knowledge, things do not become more comprehensible but more mysterious."
                                                            Albert Schweitzer

The problem with holism is its complexity. A holistic approach is methodically much harder and more challenging to grasp than a mechanistic approach. Often there are no generic conclusions or one-for-all solutions. Therefore, understanding the body as a holistic system naturally has its challenges, but let's take this idea even further. We can probably agree that the complexity of integrated systems is apparent all around us; may that be the complexity of physical systems like the human body, ecosystems, social systems, or political systems. The question is, why is a holistic approach not common practice (anymore)? And why is acceptance of all aspects of holism so challenging at times, even for those of us who recognise its trueness?

HOLISM

The term 'holism' was introduced into our language less than a hundred years ago, yet the knowledge of wholeness is thousands of years old. More than 300 years B.C., its principle was concisely summarised by Aristotle in his famous quote, "the whole is greater than the sum of its parts". Holism considers natural systems (physical, biological, chemical, social, economic, mental, linguistic, etc.) as integrated wholes, not merely assemblies of separate parts. The notion is that all properties and workings of a system cannot be determined or explained by understanding each part in isolation. The system as a whole determines how the parts behave.

The opposite of holism is reductionism. Reductionism takes the position that a complex system can be explained by cutting it down it to its fundamental parts.

Dualism

In the early 17th century, an influential thinker had three visions brought to him by what he believed to be divine spirit. They revealed that all truths were linked, and proceeding with logic to find the fundamental truth would open the way to science. Soon after, the philosopher, mathematician, and scientist René Descartes found his truth, "I exist", from which his most famous quote sprang: "cogito ergo sum". Commonly it is translated as "I think, therefore I am"; less commonly, though maybe more accurately as "I doubt, therefore I am". Descartes believed that spirit and matter are two separate substances. We have a body, and separate from it, we have a mind that interacts with the body.

This father of modern rationalism popularised what is nowadays known as Cartesian dualism (derived from his Latin name Renatus Cartesius). Like it or not, dualism has seriously influenced our way of thinking. Rationalism regards reason as the chief source and benchmark for knowledge. Therefore, truth is believed to be intellectual and deductive, not sensory and inclusive. The resulting disembodiment gradually reduced our intrinsic knowledge of holism to a more and more mechanistic way of thinking.

Separation and Integration

There is nothing either good or bad but thinking makes it so.
William Shakespeare

As fundamentally different as holism and dualism are, both aim for a better understanding of the state and character of being human. This is something to be appreciated regardless of how you think or feel about the conversation regarding holism and dualism.

Mechanistic Thinking

Mechanistic thinking proposes that humanhood can be understood by:

- Dissection
- Measurability
- Categorisation
- Causation

Dissection: Anything can be divided into small pieces to be understood, regardless of interdependence and context.

Measurability: Real things are measurable.

Categorisation: A situation is black or white, right or wrong; there is either or.

Causation: One event leads to another in a predictable manner. Said differently, cause and effect have a predictable linearity.

Holistic Thinking

Holistic thinking proposes that understanding humanhood requires consideration of the:

- Context
- Relationships
- Unmeasurable
- Variables

Context: The meaning of something changes with context.

Relationships: Everything is in relationship with other things.

Unmeasurable: Not everything that is 'real' or matters can be measured.

Variables: A sequence of events often occurs in a non-linear fashion. Between cause and effect are variables that are often unpredictable.

Understanding Wholeness in Parts

I can't see mechanistic principles to be a reflection of humanhood, but I can certainly appreciate the tangible clues derived from dissection, measurements, categorisation, and predictions. Generally, for a 'Cartesianised' mind like mine, understanding a part is easier and often necessary to get a grasp of what wholeness means. And that is okay. With awareness, the insights gained from a mechanistic approach can serve as practical stepping-stones toward a better understanding and embodiment of holism. It is imperative that mechanistic ideas are utilised as informative, thought-provoking means and not presumed to be the final truth.

Embodying the Mind Through Movement

Of course, what I called the 'problem' with holism is not really a problem. Over the past 300 years, we have just been conditioned to believe that it is. Part of the real problem is the pervasive nature of intellectual conditioning. It seems that even those of us who have been consciously reprogramming our Cartesian minds, occasionally yearn to be presented with an evidence-based formula from which we can derive straight-forward training solutions that guarantee holistic success for complex challenges. I think you can see the issue in this.

In the Western world, our common education system is fuelled by mechanistic ideals and reductionist ideas. Denial, wishful thinking, or using the right words on a business card are not going to change that. Acceptance, openness to change, and mindfulness in the process will.

I believe that turning a Cartesian mind into an embodied mind or a 'minded body' if you like, is a non-linear process, characterised by success and failure. It requires thinking, feeling, and acting in an integral manner, as well as humbly and with kindness to oneself, accepting when we temporarily default to dualism. A resource-oriented integrative movement practice such as Slings can support the process on a daily or weekly basis, with the added benefit of feeling strong and at ease in the body. Nothing to lose, only to be gained.

Integrity

"Integrity is choosing courage over comfort, it's choosing what's right over what's fun, fast or easy, and it's practising your values, not just professing them."
Brené Brown

PART 4

MOVEMENT IS GREATER THAN

Inspired by "Kunst aufräumen" by Ursus Wehrli

Parts of the Whole

THE SUM OF ITS PARTS

The body is more than aggregate of more or less functional parts.

Movement is more than the result of neuromuscular coordination with a fascial contribution.

A deliberate, fascia-focussed movement practice is more than going through the motions.

Resource-oriented integrative movement is more than a series of clever exercises.

Aesthetic movement is more than beautiful sequencing.

AFTERWORD

I remember the very first time I met Karin Gurtner. I was settling into a back-row chair of a classroom in Tempe, Arizona, where I would spend the next ten days diving deeply into the first three courses of Slings Myofascial Training – a concept that transformed my life and continues to do so today. Please note the 'back-row chair', as this was my norm in those days due to chronic pain, somatic fear, and debilitating solutions to avoid movement.

As Karin would do with each student in the class, she gracefully made her way down to my level and introduced herself with a disarming authenticity that still catches me off guard today. When Karin talks with a person, she does so with attention and intention that is unparalleled. Karin brings that same level of attention and intention to everything she does. It is unmistakably found on the pages of this book – a culmination of over a decade of personal exploration, dedicated study, somatic research, creative collaboration, and relentless commitment to learning. Karin's passion for her work is grounded in her desire to foster and encourage freedom of application in any movement modality and to share her learning generously.

It is a privilege and blessing to call Karin my mentor and friend. Inspired by her teaching and persuaded by discoveries of my own somatic truths, I have committed to a daily Slings practice that continues to fully connect me with this body that is me. I have found a wholeness and freedom that eluded me for most of my life. A combat veteran of the United States Navy, I came to this work with a lot of physical and emotional compensations. Slings work has allowed me to process and integrate these compensations into movement ease, greater health, and relief from layers of chronic pain. Ultimately, I am living a fuller life than I ever imagined possible. Needless to say, I am not sitting in back-row chairs anymore. It really is something to land in your body as a whole being for the first time at the age of 50! Thank you, body. Thank you, Slings. Thank you, Karin.

Karin has pressed into the complex nature of fascia and movement, mining deeply for the riches that are found in this resource you hold in your hands. I encourage you to keep absorbing (again and again) this wonderful work with an open mind. Karin herself embodies this open-mindedness every day with her 'ignorance-conscious' approach to learning, ever modelling humility and curiosity. Ignorance-conscious does not discount all that we might know; it simply frees us up to consider new ideas, unencumbered by external paradigms. I hope your Slings journey continues beyond this book. Allow the concepts and experiences to land as fresh wisdom. Perhaps you too will discover new possibilities in your own practice and work – if you've made it this far, I suspect you already have.

I once told Karin, "You help me be the very best version of myself". In true Karin form, with sincerity to her core, she replied, "You do the same for me Heidi". I hope that as you embrain and embody Karin's unique perspective on fascia and movement that you too discover the very best version of yourself.

Heidi Savage

Heidi is a Slings Myofascial Training Practitioner, editor for art of motion, and founder of Thriving Grace.
A graduate of the United States Naval Academy and Combat Veteran, Heidi currently serves as the Director of Fitness at RiverWoods Exeter in New Hampshire.

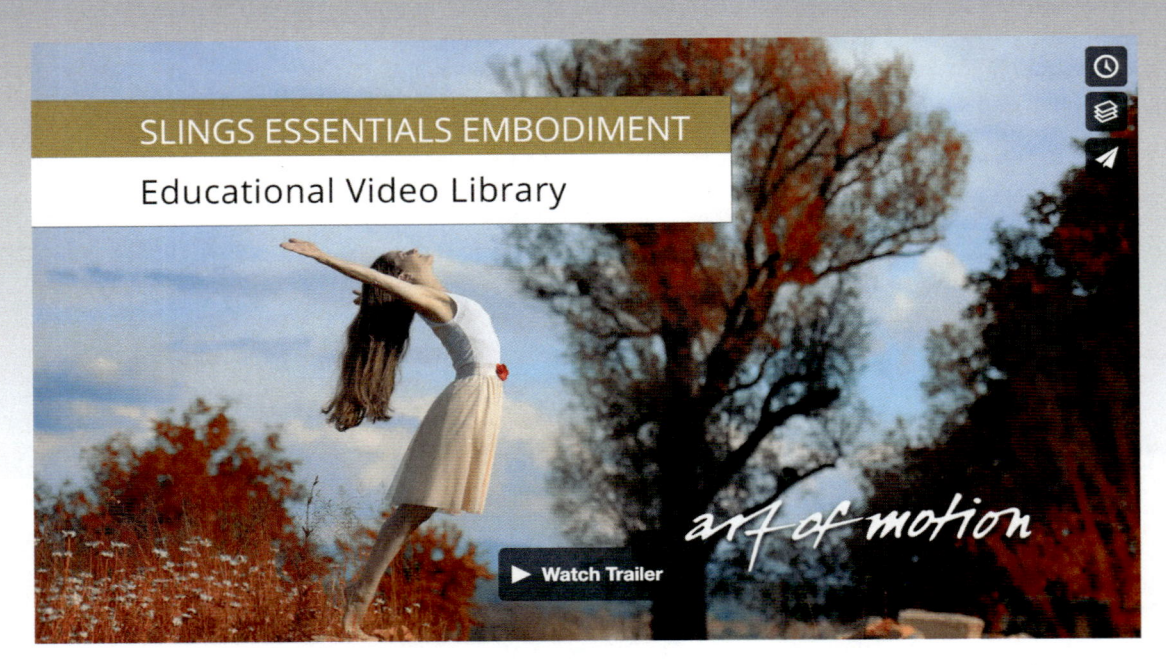

SLINGS ESSENTIALS EMBODIMENT
Educational Video Library

Slings Essentials Embodiment HD
from art of motion Academy PRO on March 21, 2020

Receive Updates

Genres: Instructional, Sports
Duration: 5 hours 43 minutes
Availability: Worldwide

SLINGS MYOFASCIAL TRAINING
A Movement Concept for Somatic Ease and Radiant Vitality
amazon.com/Anatomy-Trains-Motion-myofascial-meridian/dp/B07Y4HSTTG

Read more

1.	Slings Practice: Preparation	01:41
2.	1. Tensile Strength: Practice	26:59
3.	2. Muscle Collaboration: Practice	24:26

Buy all — Stream + download anytime

Watch on iOS, Android, Apple TV, Roku, and Chromecast. Learn more

Slings Essentials Embodiment
Educational Video Library

SLINGS ESSENTIALS VIDEO LIBRARY ON VIMEO ON DEMAND

This educational video library contains twelve fascia-focussed movement practices. A short story accompanies each practice.

Every practice-story duo focuses on one of the Fascial Movement Qualities in the Slings Essentials manual. The collection is versatile, yet each practice has its own clearly defined theme and flavour.

In its own right, the rich content of this library will feed the brain and nourish the body. However, for the extra-curious mover, body-minded movement professionals and therapists, or the fascia nerds among us, the Slings Essentials manual for embodied learning is highly recommended. Combining the manual with these practices will multiply your understanding of the 'why', 'what', and 'how' of resource-oriented, integrative movement.

If you have the Slings Essentials manual, you can use the included discount code to get a 25% reduction for this video library.

The video library is available on Vimeo on Demand: **vimeo.com/ondemand/slingsessentials**

Discount code for a 25% price reduction: **Slings-Essentials-VIP**

Practice Videos

1. Tensile Strength
2. Muscle Collaboration
3. Force Transmission
4. Adaptability
5. Multidimensionality
6. Fluidity
7. Glide
8. Elasticity
9. Plasticity
10. Tone Regulation
11. Kinaesthesia
12. Imponderability

Story Videos

1. Tensile Strength
2. Muscle Collaboration
3. Force Transmission
4. Adaptability
5. Multidimensionality
6. Fluidity
7. Glide
8. Elasticity
9. Plasticity
10. Tone Regulation
11. Kinaesthesia
12. Imponderability

Do not quote what I once said, I am wiser now

Karin staying curious in a library of her dreams in Môtier in Switzerland

REFERENCES

Baar, K. (2020): *Effect of Load and Nutrition on Connective Tissues.* From Fascia Research Society [Video recording]. Retrieved from: https://fasciaresearchsociety.org/webinar

Benias, P. C., Wells, R. G., Sackey-Aboagye, B., Klavan, H., Reidy, J., Buonocore, D., ... Theise, N. D. (2018). Structure and Distribution of an Unrecognized Interstitium in Human Tissues. *Scientific Reports, 8*(1). doi: 10.1038/s41598-018-23062-6

Bissell M. (2012). *Experiments that Point to a New Understanding of Cancer.* On TED [Video recording]. Retrieved from: https://www.ted.com/talks/mina_bissell_experiments_that_point_to_a_new_understanding_of_cancer

Blakeslee, S., Blakeslee, M. (2007). *The Body Has a Mind of Its Own: How Body Maps in Your Brain Help You Do (Almost) Everything Better.* New York: Random House.

Bochaton-Piallat, M.-L., Gabbiani, G., Hinz, B. (2016). The Myofibroblast in Wound Healing and Fibrosis: Answered and Unanswered Questions. *F1000Research,* 5, 752. doi: 10.12688/f1000research.8190.1

Bordoni, B., Zanier, E. (2015). Understanding Fibroblasts in Order to Comprehend the Osteopathic Treatment of the Fascia. *Evidence-Based Complementary and Alternative Medicine, 2015,* 1–7. doi: 10.1155/2015/860934

Calsius, J. (2020): *Interoception: A New Correlate for Intricate Connections Between Fascial Receptors, Emotions and Self-awareness.* From Fascia Research Society [Video recording]. Retrieved from: https://fasciaresearchsociety.org/webinar

Craig, A. D. (2015). *How Do You Feel? An Interoceptive Moment with Your Neurobiological Self.* Princeton, NJ: Princeton University Pres.

Damasio, A. R. (2019). *The Strange Order of Things: Life, Feeling, and the Making of Cultures.* New York: Vintage Books, a division of Penguin Random House LLC.

Davidson, R. J., Begley, S. (2012). *The Emotional Life of Your Brain: How Its Unique Patterns Affect the Way You Think, Feel, and Live – And How You Can Change Them.* London, England: Penguin Books Ltd.

Earls, J. (2014). *Born to Walk: Myofascial Efficiency and the Body in Movement.* Berkeley: North Atlantic Books.

Fede, C. (2020). *Cells of Fascia and Their Receptors.* In Fascia Research Online Summit [Video recording]. Retrieved from: https://youtu.be/_5PjQLzFbzg

Frank, G., Storch, M. (2010). *Die Mañana-Kompetenz: Entspannung als Schlüssel zum Erfolg.* München: Piper.

Gracovetsky, S. (2011). *Is the Lumbodorsal Fascia Necessary?* [Video recording]. Retrieved from: https://functionalanatomy-blog.com/2011/03/07/lecture-by-serge-gracovetsky-is-the-lumbodorsal-fascia-necessary/

Guimberteau, J. C. (2015). *Fascia Sliding.* [Video recording], Fourth International Fascia Research Congress. Washington, DC. Retrieved from: https://fasciacongress.org/dvd-recordings-and-books/

Hedley, G. (2006). *The Integral Anatomy Series, Vol. 2 Deep Fascia and Muscle* [DVD]. Melbourne, FL. Integral Anatomy Productions, LLC.

Hodges, P. (2020): *Pain, Motor Control, Fascia & Inflammation: Implications for Rehabilitation.* From Fascia Research Society [Video recording]. Retrieved from: https://fasciaresearchsociety.org/webinar

Klinger, W. (2012). *Temperature Effects of Fascia.* [Video recording], Retrieved from: https://www.youtube.com/watch?v=fz7wrf7UdjU

Kram, R., Dawson, T. J. (1998). Energetics and Biomechanics of Locomotion by Red Kangaroos (Macropus Rufus). *Comparative Biochemistry and Physiology Part B: Biochemistry and Molecular Biology, 120*(1), 41–49. doi: 10.1016/s0305-0491(98)00022-4

Langewin, H. (2020): *Effect of Stretching on the Resolution of Inflammation within Connective Tissue.* From Fascia Research Society [Video recording]. Retrieved from: https://fasciaresearchsociety.org/webinar

Li, B., Wang, J. H.-C. (2011). Fibroblasts and Myofibroblasts in Wound Healing: Force generation and measurement. *Journal of Tissue Viability, 20*(4), 108–120. doi: 10.1016/j.jtv.2009.11.004

Lipton, B. H. (2008). *The Biology of Belief: Unleashing the Power of Consciousness, Matter & Miracles.* Carlsbad, CA: Hay House.

Medina, J. (2008). *Brain Rules: 12 Principles for Surviving and Thriving at Work, Home, and School.* Seattle, WA: Pear Press.

Myers, T. W. (2012). *Dynamic Ligaments.* Retrieved from https://www.embryo.nl/upload/documents/artikelen-fascie/Dynamic Ligaments The Revolutionary Re-vision of Jaap van der Wal 2011 EN article.pdf

Myers, T. W. (2014). *Anatomy Trains. Myofascial Meridians for Manual and Movement Therapists* (3rd ed.). Philadelphia: Elsevier.

Myers, T. W. (2015). *Foam Rolling and Self-Myofascial Release*. Retrieved from: https://www.anatomytrains.com/blog/2015/04/27/foam-rolling-and-self-myofascial-release/

Myers, T. W. (2017). *BodyReading: Visual Assessment and the Anatomy Trains* (1st ed.). Walpole, ME: Anatomy Trains.

Purslow, P. P. (2010). Muscle Fascia and Force Transmission. *Journal of Bodywork and Movement Therapies, 14*(4), 411–417. doi: 10.1016/j.jbmt.2010.01.005

Schleip, R. (2003). Fascial Plasticity – A New Neurobiological Explanation Part 2. *Journal of Bodywork and Movement Therapies, 7*(2), 104–116. doi: 10.1016/s1360-8592(02)00076-1

Schleip, R., Klingler, W., Lehmann-Horn, F. (2006). Fascia is Able to Contract in a Smooth Muscle-Like Manner and Thereby Influence Musculoskeletal Mechanics. *Journal of Biomechanics, 39*. doi: 10.1016/s0021-9290(06)84993-6

Schleip, R., Chaitow, L., Findley, T., Huijing, P. (2012). *Fascia: The Tensional Network of the Human Body. The science and clinical applications in manual and movement therapy* (1st ed.). Edinburgh: Churchill Livingstone/Elsevier.

Schleip, R., (2017). *Fascia as a Sensory Organ. A Target of Manipulation*. Retrieved from: http://axissyllabus.org/assets/pdf/Schleip_Fascia_as_a_sensory_organ.pdf

Schleip, R., Gabbiani, G., Wilke, J., Naylor, I., Hinz, B., Zorn, A., … Klingler, W. (2019). Fascia Is Able to Actively Contract and May Thereby Influence Musculoskeletal Dynamics: A Histochemical and Mechanographic Investigation. *Frontiers in Physiology, 10*. doi: 10.3389/fphys.2019.00336

Schleip, R. (2020): *Elastic Storage Properties of the Fascial System*. In Fascia Research Online Summit [Video re-cording]. Retrieved from: https://youtu.be/CdGgmrW9S70

Schleip, R. (2020): *Fascia and the Autonomic Nervous System*. From Fascia Research Society [Video recording]. Retrieved from: https://fascia-researchsociety.org/webinar

Standley, P. R., Meltzer, K. (2008). In Vitro Modeling of Repetitive Motion Strain and Manual Medicine Treatments: Potential Roles for Pro- and Anti-Inflammatory Cytokines. *In Journal of Bodywork and Movement Therapies, 12*(3), 201–203. doi: 10.1016/j.jbmt.2008.05.006

Stecco, A. (2020): *Fascial Densification*. From Fascia Research Society [Video recording]. Retrieved from: https://fasciaresearchsociety.org/webinar

Stecco, A. (2015). Imaging and Measurement Technologies. In *Fourth International Fascia Research Congress* [Video recording], Washington, DC. Retrieved from: https://fasciacongress.org/dvd-recordings-and-books/

Stecco, C., Hammer, W. I. (2014). *Functional Atlas of the Human Fascial System (1st ed.)*. Edinburgh: Churchill Livingstone.

Stecco, C. (2020): *Fascial Aspects of the Human Pelvic Floor*. In Fascia Research Online Summit [Video recording]. Retrieved from: Fascial aspects of the human pelvic floor: https://youtu.be/IildXjzLVo0

Stecco, C. (2020): *Innervation of Fascia*. From Fascia Research Society [Video recording]. Retrieved from: https://fasciaresearchsociety.org/webinar

Stecco, C., Fede, C., Macchi, V., Porzionato, A., Petrelli, L., Biz, C., … Caro, R. D. (2018). The Fasciacytes: A New Cell Devoted to Fascial Gliding Regulation. *Clinical Anatomy, 31*(5), 667–676. doi: 10.1002/ca.23072

Theise N.D. (2020). *Continuity of Fibrous Tissue Interstitial Spaces Throughout the Human Body: Implications for Pathology*. From pathCast [Video recording]. Retrieved from: https://www.youtube.com/watch?v=Zd1KxVRS5Ig

Tittel, K. (2003). *Beschreibende und Funktionelle Anatomie des Menschen*. München: Urban & Fischer.

Watkins, J. (2009). *The pocket podiatry guide: functional anatomy*. Edinburgh: Churchill-Livingstone.

Wilke, J. (2020): *Myofascial Force Transmission Lines in the Human Body*. From Fascia Research Society [Video recording]. Retrieved from: https://fasciaresearchsociety.org/webinar

TENSEGRAL COLLABORATION

The creation of this book and the corresponding video library has been a tensegral collaboration at its best. Thank you!

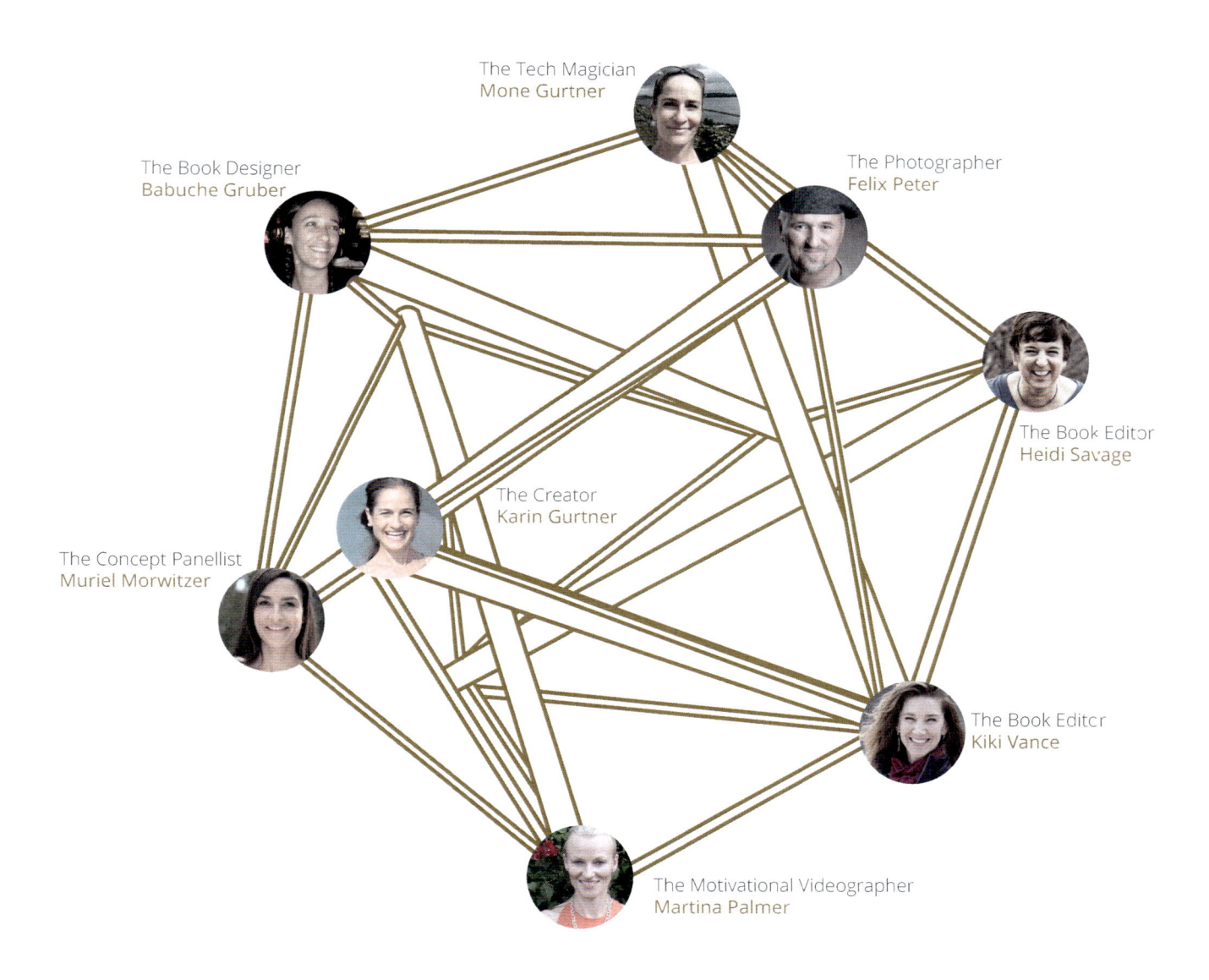

The Tech Magician
Mone Gurtner

The Photographer
Felix Peter

The Book Designer
Babuche Gruber

The Book Editor
Heidi Savage

The Concept Panellist
Muriel Morwitzer

The Creator
Karin Gurtner

The Book Editor
Kiki Vance

The Motivational Videographer
Martina Palmer

MORE GOOD THINGS TO FEED YOUR BRAIN AND NOURISH YOUR BODY

Anatomy Trains in Motion Study Guide

The study guide includes detailed maps for each myofascial meridian, listings of functional anatomy, training aims and considerations specific to each line, as well as the recommended movement sequences to embody your learning. This is a power-packed practical tool for understanding Thomas W. Myers' Anatomy Trains concept through the lens of resource-oriented, fascia-focused movement.

www.amazon.com

Slings Myofascial Training Education: Face-to-Face and Online

In collaboration with education partners around the world, art of motion Academy offers the comprehensive Slings Myofascial Training courses, including Anatomy Trains in Motion, internationally. All modules are interactive face-to-face courses. A limited portion of the curriculum is available in the form of in-depth online learning. Formats can be mixed and matched to suit the learner.

www.art-of-motion.com

Daily Slings Practice

From experience, we know that a healthy dose of Slings each day, or each week, enhances life quality on many levels. Therefore, we offer a versatile collection of functionally choreographed Slings practices on Vimeo. The short lessons are designed for people who want to nourish their bodies with wholesome exercise, as well as movement professionals looking for teaching inspiration.

www.art-of-motion.com/en/shop

Slings Myofascial Training® Tools

Made with love, the Slings props were designed with fascia in mind.

- 1 Kneeling Pad, 50 x 24 x 1,5 cm, anthracite
- 2 Slings Myofascial Massage Balls, 10 cm Ø, 100 g, gold
- 2 Slings Myofascial Massage Domes, 11 cm Ø, 6 cm height, 150 g, anthracite
- 2 Slings Myofascial Trigger Balls, 40 mm Ø, 20 g, silver
- 1 Bag

Please visit the art of motion online shop for the list of international distributors.

www.art-of-motion.com/en/shop

"Intelligent exercise is a commitment to wholeness;
enhancing and enjoying our full humanity while we can."
Damon Young

Made in the USA
Coppell, TX
16 May 2021